my revision notes

AQA GCSE (9–1)

GEOGRAPHY

Simon Ross
Rebecca Blackshaw

HODDER
EDUCATION
AN HACHETTE UK COMPANY

Publishers would like to thank the following for permission to reproduce copyright material.

Photo credits

p.3 *t, c & b* © Henning Dalhoff/Science Photo Library; **p.4** © Tommy E Trenchard/ Alamy Stock Photo; **p.10** © NOAA (National Oceanic and Atmospheric Administration); **p.11** *l* © ODD ANDERSEN/AFP/Getty Images, *r* © NOEL CELIS/AFP/Getty Images; **p.12** © Simon Clarke, ShelterBox; **p.23** © Getty Images/ iStockphoto/Thinkstock; **p.28** © Getty Images/iStockphoto/Thinkstock; **p.32** © Worldwide Picture Library/Alamy Stock Photo; **p.49** © REUTERS/Alamy Stock Photo; **p.78** © Simon Ross; **p.85** © Full On Adventure/https://www.flickr.com/ photos/fullonadventure/4297947326/https://creativecommons.org/licenses/by/2.0/; **p.87** © Fix the Fells; **p.96** © Novarc Images/Alamy Stock Photo; **p.97** © Getty Images/ iStockphoto/Thinkstock; **p.118** *l & r* © Getty Images/iStockphoto/Thinkstock; **p.129** © Geoffrey Robinson/Alamy Stock Photo; **p.140** © Milleflore Images – Holidays Events/ Alamy Stock Photo; **p.141** © Martin Parker/Alamy Stock Photo; **p.172** *l & r* Simon Ross.

Acknowledgements

Every effort has been made to trace all copyright holders, but if any have been inadvertently overlooked, the Publishers will be pleased to make the necessary arrangements at the first opportunity.

Although every effort has been made to ensure that website addresses are correct at time of going to press, Hodder Education cannot be held responsible for the content of any website mentioned in this book. It is sometimes possible to find a relocated web page by typing in the address of the home page for a website in the URL window of your browser.

Hachette UK's policy is to use papers that are natural, renewable and recyclable products and made from wood grown in sustainable forests. The logging and manufacturing processes are expected to conform to the environmental regulations of the country of origin.

Orders: please contact Bookpoint Ltd, 130 Park Drive, Milton Park, Abingdon, Oxon OX14 4SE. Telephone: (44) 01235 827720. Fax: (44) 01235 400454. Email education@bookpoint.co.uk Lines are open from 9 a.m. to 5 p.m., Monday to Saturday, with a 24-hour message answering service. You can also order through our website: www.hoddereducation.co.uk

ISBN: 9781471887314

© Simon Ross and Rebecca Blackshaw 2017

First published in 2017 by

Hodder Education,
An Hachette UK Company
Carmelite House
50 Victoria Embankment
London EC4Y 0DZ

www.hoddereducation.co.uk

Impression number   10   9   8   7   6   5   4   3

Year   2021   2020   2019   2018

Cover photo © Frederic Bos - Fotolia.com
Illustrations by Barking Dog Art
Typeset in Integra Software Services Pvt. Ltd., Pondicherry, India
Printed in Spain

A catalogue record for this title is available from the British Library.

Get the most from this book

Everyone has to decide his or her own revision strategy, but it is essential to review your work, learn it and test your understanding. These Revision Notes will help you to do that in a planned way, topic by topic. Use this book as the cornerstone of your revision and don't hesitate to write in it – personalise your notes and check your progress by ticking off each section as you revise.

Tick to track your progress

Use the revision planner on pages iv–vi to plan your revision, topic by topic. Tick each box when you have:

- revised and understood a topic
- tested yourself
- practised the exam questions and gone online to check your answers

You can also keep track of your revision by ticking off each topic heading in the book. You may find it helpful to add your own notes as you work through each topic.

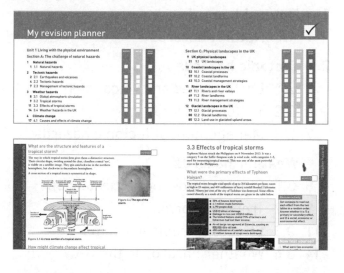

Features to help you succeed

Exam tips

Expert tips are given throughout the book to help you polish your exam technique in order to maximise your chances in the exam.

Now test yourself

These short, knowledge-based questions provide the first step in testing your learning. Answers can be found online at: **www.hoddereducation.co.uk/myrevisionnotes**

Key terms

Key terms are highlighted in the text. You can find the definitions of these words in the subject specific vocabulary PDF on the AQA website www.aqa.org.uk/subjects/geography/gcse/geography-8035/teaching-resources.

Revision activities

These activities have been designed to focus your revision and will help you to understand each topic in an interactive way.

Exam practice

Practice exam questions are provided for each topic. Use them to consolidate your revision and practise your exam skills.

Online

Go online to check your answers to the exam questions at **www.hoddereducation.co.uk/myrevisionnotes**

Case studies and examples

Revision notes on case studies and examples are included so that you can give specific details in your answers. Sometimes more than one location has been covered but **you only need to remember the details for the one that you have studied in class**.

If you have studied a different location from what is covered in this book, then you can use your own notes to revise.

My revision planner

REVISED   TESTED   EXAM READY

Section C: Physical landscapes in the UK

REVISED   TESTED   EXAM READY

Unit 2 Challenges in the human environment

Section A: Urban issues and challenges

REVISED   TESTED   EXAM READY

Section B: The changing economic world

REVISED   TESTED   EXAM READY

REVISED   EXAM READY

Countdown to my exams

6–8 weeks to go

- Start by looking at the specification – make sure you know exactly what material you need to revise and the style of the examination. Use the revision planner on pages iv and vi to familiarise yourself with the topics.
- Organise your notes, making sure you have covered everything on the specification. The revision planner will help you to group your notes into topics.
- Work out a realistic revision plan that will allow you time for relaxation. Set aside days and times for all the subjects that you need to study, and stick to your timetable.
- Set yourself sensible targets. Break your revision down into focused sessions of around 40 minutes, divided by breaks. These Revision Notes organise the basic facts into short, memorable sections to make revising easier.

REVISED ☐

2–6 weeks to go

- Read through the relevant sections of this book and refer to the exam tips, and key terms. Tick off the topics as you feel confident about them. Highlight those topics you find difficult and look at them again in detail.
- Test your understanding of each topic by working through the 'Now test yourself' questions in the book. Look up the answers online.
- Make a note of any problem areas as you revise, and ask your teacher to go over these in class.
- Look at past papers. They are one of the best ways to revise and practise your exam skills. Write or prepare planned answers to the exam practice questions provided in this book. Check your answers online at **www.therevisionbutton.co.uk/myrevisionnotes**
- Try out different revision methods. For example, you can make notes using mind maps, spider diagrams or flash cards.
- Track your progress using the revision planner and give yourself a reward when you have achieved your target.

REVISED ☐

One week to go

- Try to fit in at least one more timed practice of an entire past paper and seek feedback from your teacher, comparing your work closely with the mark scheme.
- Check the revision planner to make sure you haven't missed out any topics. Brush up on any areas of difficulty by talking them over with a friend or getting help from your teacher.
- Attend any revision classes put on by your teacher. Remember, he or she is an expert at preparing people for examinations.

REVISED ☐

The day before the examination

- Flick through these Revision Notes for useful reminders, for example the exam tips and key terms.
- Check the time and place of your examination.
- Make sure you have everything you need – extra pens and pencils, tissues, a watch, bottled water, sweets.
- Allow some time to relax and have an early night to ensure you are fresh and alert for the examinations.

REVISED ☐

My exams

Paper 1: Living with the physical environment

Date:..

Time:..

Location:..

Paper 2: Challenges in the human environment

Date:..

Time:..

Location:..

Paper 3: Geographical applications

Date:..

Time:..

Location:..

1 Natural hazards

1.1 Natural hazards

A **natural hazard** is a natural event such as an **earthquake**, volcanic eruption, **tropical storm** or flood. It is caused by 'natural' processes, so would occur even if humans were not on the planet. However, it is a 'hazard' because it has the *potential* to cause damage, destruction and death when it interacts with humans.

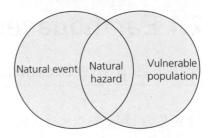

Figure 1.1 A natural hazard

REVISED

What are types of natural hazard?

Natural hazards are most commonly classified by the physical processes that caused them:

- Earthquakes and **volcanoes** are classed as **tectonic hazards**.
- Flooding is a geomorphological hazard.
- Tropical storms are classed as atmospheric hazards.
- Forest fires are biological hazards.

However, classifying hazards is made difficult because:

- some may be caused by more than one physical process; for example, a tsunami can be both a tectonic **and** a geomorphological hazard
- some may also be caused or influenced by human processes; for example, forest fires can be a natural biological hazard but can also be caused by human activity such as arson or falling power lines.

Factors affecting hazard risk

Hazard risk is the probability or chance that a natural hazard may take place. The hazard risk will increase when there is:

1. an increase in the number of people vulnerable to the natural hazard; for example:
   - world population has increased
   - more people are living near hazard-prone areas because they cannot move, it's worth staying or simply they don't want to move
2. an increase in the a) frequency (how often) and b) magnitude (strength) of the natural hazard; for example:
   - some hazards are more destructive than others
   - global warming increases the probability of hazards such as flooding
   - **deforestation** and urbanisation increase hazards such as landslides and flooding.
3. a decrease in the number of people capable of coping with the natural hazard; for example:
   - some hazards are harder to predict and it is therefore harder to evacuate people in time
   - some people do not have the money, knowledge or ability to cope with natural hazards when they occur. This is more likely in **low income countries (LICs)** than in **high income countries (HICs)**.

> **Revision activity**
>
> List as many different natural hazards as you can, then colour code them to show what type of natural hazard they can be classified as.

> **Exam practice**
>
> 1. Define natural hazard. (1 mark)
> 2. Suggest how hazard risk would be affected by an increase in population. (2 marks)
>
> ONLINE

> **Now test yourself**
>
> 1. What is hazard risk?
> 2. List at least three factors which affect hazard risk.
>
> TESTED

2 Tectonic hazards

2.1 Earthquakes and volcanoes

What is plate tectonics theory?

REVISED

The Earth's structure

The Earth's internal structure is divided into layers; the core, mantle and crust (continental and oceanic). The crust and upper mantle are called the lithosphere. The lithosphere is broken into several major fragments called **tectonic plates**. Tectonic plates are rigid and can move very slowly, floating across the heavier semi-molten rock in the mantle. Continental plates are less dense, but thicker than oceanic plates.

What causes tectonic plates to move?

1 One theory is called **convection**.
2 A more currently accepted theory is **ridge push and slab pull**.

> **Revision activity**
>
> 1 Draw a labelled diagram to show the Earth's structure.
> 2 Time yourself: write out, in four minutes, why tectonic plates move.
> 3 After reading page 3, draw three simple sketches of the different plate margins. Annotate the sketches with the information in the table.

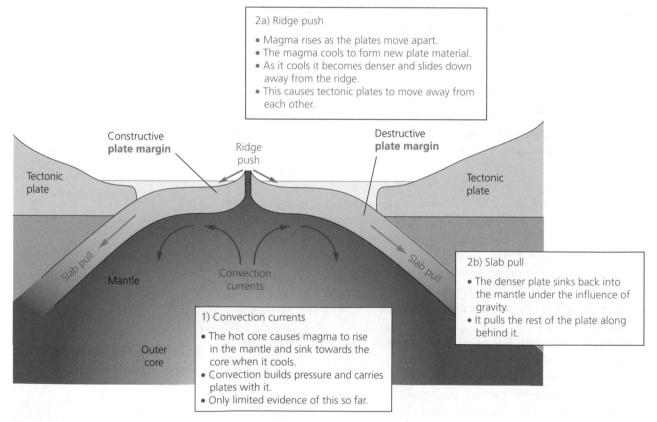

2a) Ridge push

- Magma rises as the plates move apart.
- The magma cools to form new plate material.
- As it cools it becomes denser and slides down away from the ridge.
- This causes tectonic plates to move away from each other.

Constructive **plate margin** Ridge push Destructive **plate margin**

Tectonic plate Tectonic plate

Slab pull Slab pull

Mantle Convection currents

1) Convection currents

- The hot core causes magma to rise in the mantle and sink towards the core when it cools.
- Convection builds pressure and carries plates with it.
- Only limited evidence of this so far.

2b) Slab pull

- The denser plate sinks back into the mantle under the influence of gravity.
- It pulls the rest of the plate along behind it.

Outer core

Figure 2.1 Plate tectonics theory

What is the global distribution of earthquakes and volcanoes in relation to plate margins?

- The distribution is not random.
- They occur in narrow bands along plate margins.
- Found both on land and in sea.
- Earthquakes found at all three types of **plate margins**: **constructive**, **destructive** and **conservative**.
- Volcanoes found at constructive and destructive plate margins.
- There are anomalies as some occur in the middle of plates in 'hot spots'.

Now test yourself

1 Which plate margins experience a) earthquakes and b) volcanoes?
2 How are the plates moving at a) constructive, b) destructive and c) conservative plate margins?

TESTED

What are the physical processes at different plate margins?

There are three types of plate margin: constructive, destructive and conservative.

Plate margin	Direction of plate movement	Physical process	Earthquakes	Volcanic eruptions
Constructive	Diverging away from each other, e.g. the Eurasian and North American plates	• Hot molten magma rises between the plates. • Tectonic plates move away from each other by ridge push and slab pull. • The magma cools to form new plate. • On land rift valleys form, such as the East African rift valley.	Yes (usually small, not violent)	Yes (shield)
Destructive	Converging towards each other, e.g. the Pacific and Philippine plates	• When tectonic plates converge, pressure builds between them. The rock eventually fractures, causing earthquakes. • When oceanic and continental plates collide, the denser oceanic plate is subducted under the continental plate into the mantle, where it melts. • Hot magma can rise through the lithosphere and erupt as lava through volcanoes.	Yes (violent)	Yes (composite)
Conservative	Sliding parallel past each other, e.g. Pacific and North American plates	• Pressure builds at the margin of the tectonic plates as they are pulled along behind a plate being subducted elsewhere (slab pull). • As friction is overcome, the rock fractures in an earthquake.	Yes	No

Exam practice

1 Describe the global distribution of earthquakes and volcanoes in relation to plate margins. (3 marks)
2 Outline differences between conservative and destructive plate margins. (2 marks)
3 Explain how volcanoes occur at destructive plate margins. (4 marks)
4 Draw an annotated diagram to explain why earthquakes occur at conservative plate margins. (4 marks)

ONLINE

Exam tip

To describe distributions of data given on a map, give the general overall trend, some specific examples (remembering to use map directions), and identify any anomalies.

2.2 Tectonic hazards

What are the effects of tectonic hazards?

There are two types of effects: **primary effects** and **secondary effects**.

Primary effects of an earthquake

- Property, buildings and homes destroyed.
- People injured and killed.
- Ports, bridges, roads and railways damaged.
- Pipes (water and gas) and electric cables broken.

Secondary effects of an earthquake

- Business reduced and money spent repairing damage, so the economy slows.
- Blocked transport infrastructure hinders emergency services, causing further casualties.
- Broken gas pipes and fallen electricity cables can start fires, further destroying property and killing people.
- Burst water pipes lead to a lack of clean water and poor sanitation, increasing the spread of diseases.

Businesses damaged so local economy damaged, unemployment and reduced income.

Electricity cables damaged.

Buildings and homes destroyed, injuring and killing people.

Transport infrastructure damaged. Blocked roads hinder access for emergency services and trade.

Figure 2.2 Haiti earthquake, 2010

> **Exam tip**
>
> As shown in the photograph, always make sure the arrows for your annotations touch the exact point you are referring to.

Primary effects of a volcanic eruption

- Property and farm land destroyed. People and livestock injured and killed. This is due to pyroclastic and lava flows and ash collapsing buildings.
- Air travel halted due to airborne volcanic ash damaging engines.
- Water supplies contaminated.

Secondary effects of a volcanic eruption

- Economy slows. Emergency services struggle to arrive.
- Ice melts, causing flooding. Flood water or rain mixes with volcanic ash, causing lahars (mudflows), destroying property and killing people.
- Tourism increases with those interested in visiting volcanoes.
- The ash breaks down, forming fertile farm land.

Now test yourself

1 What is the difference between a primary and a secondary effect?
2 Which effects in the annotated photograph of Haiti are a) primary and b) secondary?
3 What are the positive effects of tectonic hazards?

TESTED

What are the responses to tectonic hazards?

Responses can be **immediate** or **long-term**.

Immediate	Long-term
● Issue warnings (volcanic eruptions if possible). ● Rescue teams search for survivors. ● Treatment given to those injured. ● Provide shelter, food and drink. ● Recover bodies. ● Extinguish fires.	● Repair and rebuild properties and transport infrastructure. ● Improve building regulations. ● Restore utilities (water, gas, electricity). ● Resettle locals elsewhere (Montserrat still has an exclusion zone over two decades later). ● Develop opportunities for recovery of the economy. ● Install increased monitoring technology.

Comparing earthquakes

You are required to study named examples of a tectonic hazard to show how the effects and responses vary depending on their contrasting levels of wealth. Italy and Chile are higher income countries (HICs) with a gross national income (GNI) per capita (2015) of US$32,790 and US$14,060 respectively, whereas Nepal is a lower income country (LIC) with a gross national income (GNI) per capita of US$730.

Varying effects and responses

	L'Aquila, Italy (2009)	Maule, Chile (2010)	Gorka, Nepal (2015)
Effects	**Primary** ● 308 killed and 1,500 injured. ● US$11,434 million damage. ● 67,500 homeless. ● 10,000–15,000 buildings collapsed, including National Museum, Porta Napoli, several L'Aquila University buildings and San Salvatore Hospital. **Secondary** ● Landslides and rockfalls damaged housing and transport. ● Mudflow caused by a burst water pipeline near Paganio. ● House prices and rents increased. ● Some of city centre cordoned off, reducing business.	**Primary** ● 500 killed and 12,000 injured. ● Around US$30 billion damage. ● 220,000 homes, 4,500 schools, 56 hospitals and 53 ports destroyed, including Port of Tacahuanao. ● Santiago airport severely damaged. **Secondary** ● 1,500 kilometres of roads damaged, mainly by landslides. ● Several coastal towns devastated by tsunami waves. ● A fire at a chemical plant near Santiago.	**Primary** ● 8,841 died and 16,800 injured. ● US$5 billion damage. ● 1 million homeless. ● 7,000 schools, 26 hospitals and Dharahara Tower (UNESCO World Heritage Site) destroyed. ● International airport congested as aid arrived. **Secondary** ● Landslides and avalanches killed nineteen on Mount Everest. Landslide blocked the Kali Gandaki River so people evacuated in case of flooding. Blocked roads slowed aid. ● Tourism employment and income declined. ● Rice seed ruined, causing food shortages and income loss.

	L'Aquila, Italy (2009)	Maule, Chile (2010)	Gorka, Nepal (2015)
Responses	**Immediate** ● 10,000 sheltered at hotels and 40,000 tents given. ● Seven dog units searched for survivors. ● Mortgages and bills for Sky TV were suspended. ● Free mobile calls. ● US$552.9 million raised from EU Solidarity Fund. ● Disasters Emergency Committee provided no aid. **Long-term** ● Torch-lit procession with Catholic Mass each year. ● No taxes for residents during 2010. ● Students' university fees were waived, and they received free public transport. ● In 2014 the 'guilty' verdict of manslaughter in 2012 for six scientists was overturned.	**Immediate** ● Emergency services arrived quickly. ● Within 24 hours temporary repairs made to the Route 5 north–south highway. ● 90% of homes had power and water restored within ten days. ● US$60 million raised. **Long-term** ● Government housing reconstruction plan to help 200,000 households launched one month later. ● Potential for economy to be rebuilt without foreign aid. ● Possibly up to four years to recover from damage to buildings and ports.	**Immediate** ● International help requested. ● Rescues from avalanches on Mount Everest made by helicopter. ● 500,000 tents provided. ● Field hospitals set up. The UN and World Health Organisation sent medical supplies to worst affected districts. ● Facebook launched a safety feature for users to indicate they're safe. Free telephone calls. **Long-term** ● US$274 million aid money committed to recovery. ● Lakes and river valleys cleared of landslide material to avoid flooding. ● Stricter building controls enforced. ● New trekking routes on Everest opened by August 2015 and permits extended by two years.

Reasons for variations

The table below shows how effects and responses vary depending on certain factors:

Reasons for variations	Causes differences in:		Influenced by wealth
	Effects	**Responses**	
Building density	✓		
Construction standards	✓		✓
Corruption		✓	✓
Hazard-prone area	✓		
Magnitude or scale	✓		
Monitoring/prediction		✓	✓
Medical facilities		✓	✓
Population density	✓		
Resources/finance		✓	✓
Secondary effects (e.g. tsunamis)	✓	✓	
Time of day/year	✓		
Trained emergency services		✓	✓
Transport infrastructure		✓	✓
Type of plate margin	✓		

Exam practice

To what extent does a country's ability to cope with the effects of a tectonic hazard depend on its wealth? Use examples from countries with contrasting wealth to support your answer. (9 marks)

ONLINE

Exam tip

Use phrases such as 'to a greater extent' and 'to a lesser extent' to answer 'To what extent' questions. Make sure you come to a conclusion.

Now test yourself

TESTED

1 Contrast the wealth of Italy or Chile with that of Nepal.
2 Identify two differences and similarities between the a) effects and b) responses of earthquakes.

2.3 Management of tectonic hazards

Why do people live at risk from tectonic hazards?

REVISED

Eight per cent of the world's population live near volcanoes. Fifty per cent of the population of the United States of America live in earthquake-prone areas.

Economic reasons for living at risk	Social reasons for living at risk
● **Geothermal energy** provides energy for the area. ● Farming the nutrient-rich soils helps agriculture. ● Mining provides energy and income. ● Tourism creates jobs and provides income. ● It may be cheaper to stay than to move.	● People want to stay near friends and family. ● The threat may not be great enough, or people don't understand the risk. ● People are confident that the buildings will keep them safe.

How can the risks from tectonic hazards be reduced?

REVISED

Volcanologists and seismologists use **monitoring**, **prediction**, **protection** and **planning** to aim to reduce the death and destruction caused by earthquakes and volcanic eruptions.

Revision activity

1 Draw a spider diagram with four legs: monitoring, prediction, protection, planning.
2 Using two different-coloured pens (one for earthquakes and one for volcanic eruptions), list different methods on each leg.

Technique	Volcanic eruption	Earthquake
Monitoring	● Changes in shape of ground and volcano using tiltmeters and Global Positioning System (GPS) satellites. ● Earthquakes near magma chamber using seismometers. ● Ground surface and river temperatures using thermal heat sensors. ● Radon and sulphur gas using gas-trapping bottles.	More difficult to monitor than volcanoes, but these can be monitored: ● Foreshocks using seismometers and GPS. ● Radon using radon detection devices.
Prediction	● Easier to predict than earthquakes as they usually give advance warning signals before erupting.	● Extremely difficult to predict time, date or exact location.
Protection	● Buildings cannot be designed to completely protect against eruption impacts. ● Evacuation instructed by authorities.	● Building and transport infrastructure design (e.g. foundations with rubber shock absorbers), although this is expensive. ● Sea walls in case of tsunamis.
Planning	● Evacuation. ● Exclusion zones. ● Education to know what to do. ● First-aid training.	● Practice drills. ● Preparing emergency supplies and location of evacuation centres. ● Securing objects/furniture.

Exam practice

1 Describe how protection reduces the damage caused by an earthquake. (4 marks)
2 Explain why reducing the risks from tectonic hazards is challenging. (4 marks)

ONLINE

Now test yourself

1 Give examples of how people can plan for earthquakes and volcanic eruptions.
2 Name four reasons why people live near volcanoes.

TESTED

3 Weather hazards

3.1 Global atmospheric circulation

Global atmospheric circulation helps to explain the location of world climate zones and the distribution of weather hazards.

What is the influence of latitude?

When the Sun's rays strike the Earth, they are concentrated differently on areas of land depending on latitude. At the Equator the Sun's rays are concentrated so it is much hotter than at the Poles where the rays are more spread out.

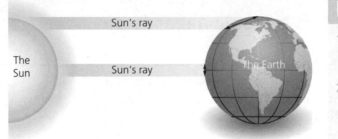

Figure 3.1 How the Sun's rays hit the Earth

Now test yourself

1 Why does air rise at the Equator and sink at the Poles?
2 What global patterns do you notice about surface winds?

TESTED

How does global atmospheric circulation work?

Pressure belts

- Air at the Equator is heated strongly so it rises in **low pressure** conditions. The air flows towards the North and South Poles. As warm air rises it cools and condenses. Low pressure therefore brings cloud and rain.
- The air sinks at 30 degrees north and south of the Equator under **high pressure**. High-pressure weather brings dry and clear skies. This forms a convection cell known as the Hadley Cell.
- Air at the **polar** latitudes is colder and denser and so the air sinks towards the ground surface under **high pressure** conditions. The air flows towards the Equator. The air warms as it reaches about 60 degrees and again rises under **low pressure** conditions. This forms the Polar Cell. Located between the Hadley and Polar Cell is the Ferrel Cell.

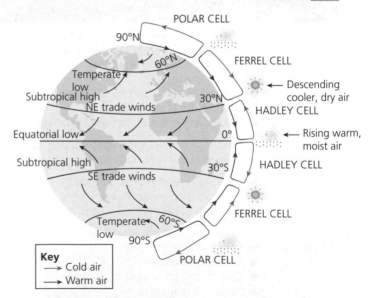

Figure 3.2 The global circulation system

Surface winds

Winds on the surface of the Earth are experienced as air moves from high to low pressure areas in the convection cells. On the surface of the Earth these winds bend due to the Coriolis effect as the Earth spins. The surface winds bend to the right in the northern hemisphere and to the left in the southern hemisphere.

Exam practice

1 Describe the weather experienced under low- pressure conditions. (2 marks)
2 Account for the apparent association between surface winds and atmospheric pressure. (4 marks)

ONLINE

3.2 Tropical storms

What is the global distribution of tropical storms?   REVISED

Tropical storms are a natural hazard. They have different names depending on their location. They occur between 5 and 30 degrees north and south of the Equator. This provides areas of intense low pressure (see previous page) so that warm, moist air is able to rise rapidly to reach high altitudes.

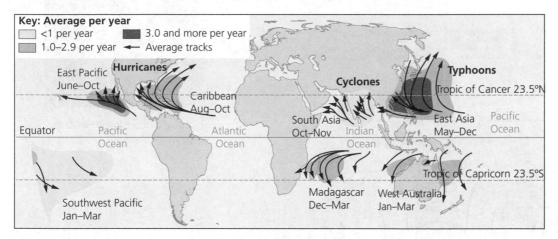

Figure 3.3 **The global distribution of tropical storms**

Conditions which cause tropical storms	How does this contribute to their formation?
Low latitudes Between 5–30 degrees north and south of the Equator	Temperatures are higher here than at the Poles so the sea and air are heated more quickly, to higher temperatures Air pressure is low, and air rises. The Coriolis effect is strong enough for tropical storms to spin.
Originate in oceans with temperatures above 27°C Ocean depth 60–70m	Provides heat and moisture so warm air rises rapidly.
Between summer and autumn	Typically the warmest seasons to encourage warmer air to rise rapidly, on account of low pressure.
Low wind shear	Wind is constant and doesn't vary with height so clouds rise to high altitudes without being torn apart.

What is the sequence of the formation of tropical storms?   REVISED

1 Air is heated above the surface of warm tropical oceans.
2 Warm air rises rapidly under low-pressure conditions.
3 Strong winds form as rising air draws up more air and moisture.
4 The rising air spins around a calm central eye of the storm due to the Coriolis effect.
5 The rising air cools and condenses, forming large cumulonimbus clouds and torrential rainfall.
6 Heat is given off as it cools, powering the tropical storm.
7 Cold air sinks in the eye so it is clear, dry and calmer.
8 The tropical storm travels across the ocean with the prevailing wind.
9 On meeting land, it loses its source of heat and moisture so loses power. Storms track north in the northern hemisphere and south in the southern hemisphere.

> **Exam tip**
>
> When describing formations, try to make your answer as sequential as possible rather than jumping around. The order is important in a clear description.

What are the structure and features of a tropical storm?

The way in which tropical storms form gives them a distinctive structure. Their circular shape, swirling around the clear, cloudless central 'eye', is visible on a satellite image. They spin anticlockwise in the northern hemisphere, but clockwise in the southern hemisphere.

A cross section of a tropical storm is symmetrical in shape.

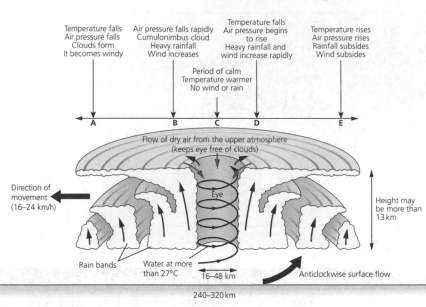

Figure 3.4 The eye of the storm

Figure 3.5 A cross section of a tropical storm

How might climate change affect tropical storms?

Climate change is expected to increase atmosphere and sea surface temperatures, and affect tropical storms in the following ways:

- **Distribution**: The location of tropical storms is not expected to change significantly, but there may be more in areas such as the South Atlantic and parts of the subtropics as sea surface temperatures increase.
- **Frequency:** The overall frequency of tropical storms is expected to remain the same or decrease. However, the frequency of category 4 and 5 storms is expected to increase, while category 1–3 storms will decrease.
- **Intensity:** Since 1970s the number of the most severe category 4 or 5 tropical storms has increased. Every 1 degree Celsius increase in sea surface temperatures will mean a 3–5 per cent increase in wind speed.

Exam tip

Circle any plurals in the exam questions. This will help ensure you notice when you need to consider more than one factor. Full marks cannot be gained unless you have obeyed this in questions.

Now test yourself

TESTED

1 Where are tropical storms called a) hurricanes, b) typhoons and c) cyclones?
2 What type of air pressure do tropical storms form in?
3 What is the typical height and width of a tropical storm?
4 What speed does a typical tropical storm travel at?
5 How do rainfall and wind change as a tropical storm passes over?
6 Name the conditions which cause tropical storms to form and which climate change may affect.

Exam practice

1 Outline two conditions in which tropical storms form. (2 marks)
2 Describe the formation of a tropical storm. (6 marks)

ONLINE

3.3 Effects of tropical storms

Typhoon Haiyan struck the Philippines on 8 November 2013. It was a category 5 on the Saffir–Simpson scale (a wind scale, with categories 1–5, used for measuring tropical storms). This was one of the most powerful ever to hit the Philippines.

What were the primary effects of Typhoon Haiyan?

REVISED

The tropical storm brought wind speeds of up to 314 kilometres per hour, waves as high as 15 metres, and 400 millimetres of heavy rainfall flooded 1 kilometre inland. Ninety per cent of the city of Tacloban was destroyed. Some effects caused directly as a result of the tropical storm are given in the table below.

Social	• 50% of houses destroyed. • 4.1 million made homeless. • 6,190 people died.
Economic	• US$12 billion of damage. • Damage to rice cost US$53 million. • The United Nations stated 75% of farmers and fishermen had lost their income.
Environmental	• An oil barge ran aground at Estancia, causing an 800,000-litre oil leak. • 400 millimetres of rainfall caused flooding. • 1.1 million tonnes of crops were destroyed.

Revision activity

Get someone to read out each effect from the two tables in a random order. Answer whether it is 1) a primary or secondary effect, and 2) a social, economic or environmental effect.

Figure 3.6 Primary effects: Tacloban city following Typhoon Haiyan, 2013

Figure 3.7 Secondary effects: residents loot water-damaged sacks of rice from a rice warehouse in Tacloban

Now test yourself

1 What were two economic impacts of Typhoon Haiyan?
2 What were two environmental impacts of Typhoon Haiyan?
3 What were two social impacts of Typhoon Haiyan?

TESTED

What were the secondary effects of Typhoon Haiyan?

REVISED

The after-effects caused indirectly by the tropical storm are listed in the table below.

Social	• Infection and diseases spread, due to contaminated surface and groundwater. • Eight deaths in a stampede as survivors fought for rice supplies. • Power supplies were cut off for a month in some areas. • Many schools were destroyed, affecting people's education.
Economic	• The fishing industry was disrupted as the leaked oil from the grounded barge contaminated fishing waters. • The airport was badly damaged and roads were blocked by trees and debris. • Looting was rife, due to a lack of food and supplies. • By 2014, rice prices had risen by nearly 12%.
Environmental	• Ten hectares of mangroves were contaminated by the oil barge leak. • Flooding caused landslides.

What was the response to Typhoon Haiyan?

The Philippine government, charities and non-governmental organisations (NGOs) responded in varying ways to reduce the effects of Typhoon Haiyan. **Immediate responses** occurred as the disaster happened, and **long-term responses** followed in the following weeks, months and years.

Immediate responses to Typhoon Haiyan

- The government televised a warning for people to prepare and evacuate.
- Authorities evacuated 800,000 people. Many went to Tacloban indoor stadium, which had a reinforced roof to withstand typhoon winds; however, unfortunately it flooded.
- Over 1,200 evacuation centres were set up to help the homeless.
- The Philippine government ensured essential equipment and medical supplies were sent out, but in one region medical supplies and equipment were washed away.
- Emergency aid supplies arrived three days later by plane. Within two weeks, over 1 million food packs and 250,000 litres of water were distributed.
- The government imposed a curfew two days after the typhoon to reduce looting.
- The celebrity couple the Beckhams, the *X Factor* TV show, and brands such as Coca-Cola, FIFA and Apple used their status to raise awareness and encourage public donations.

Figure 3.8 Humanitarian aid workers distribute ShelterBox emergency aid supplies

Long-term responses to Typhoon Haiyan

- Thirty-three countries and international organisations pledged help. More than US$1.5 billion was pledged in foreign aid.
- A 'cash for work' programme paid people to clear debris and rebuild the city.
- Oxfam replaced fishing boats.
- In July 2014, the Philippine government declared a long-term recovery plan 'Build Back Better'. Buildings would not just be rebuilt, but upgraded to protect against future disasters.
- A 'no-build zone' was established in the Eastern Visayas. Homes were rebuilt away from flood-risk areas.
- Mangroves (saltwater-adapted trees or shrubs) were replanted.
- A new storm surge warning system was installed.
- More cyclone shelters were built.

How can the effects of tropical storms be reduced?

There are four **management strategies** to cope with tropical storms:

Monitoring

- Satellites monitor cloud patterns associated with tropical storms.
- The Global Precipitation Measurement satellite monitors high-altitude rainclouds every three hours, which indicate weather a tropical storm will intensify within 24 hours.
- The National Aeronautics and Space Administration (NASA) monitors weather patterns across the Atlantic in two unmanned aircraft called Global Hawk drones.

Prediction

- Supercomputers give five days' warning and predict a location within 400 kilometres.
- Track forecast cones plot the tropical storm's predicted path. Approximately 70 per cent occur within the cone.
- Early warnings are issued by national hurricane centres around the world.

Protection

- Reinforce buildings (Figure 3.9 shows how damage to a house can be mitigated, i.e. prevented).
- Develop coastal flood defences.
- Create 'no-build zones' in low-lying areas.

Planning

Those who still live in tropical-storm-prone areas can make plans and prepare what they require to deal with the effects of the tropical storm (see also Figure 3.9). They can: prepare disaster supply kits, ensure vehicles are fuelled, know where evacuation shelters are and plan what their family will do.

Prepare disaster supplies to include medicines, non-perishable food and water for three days.

Place valuables and important paperwork in a waterproof container on the highest level.

Cover windows and doors with hurricane shutters or pre-cut marine plywood.

Set refrigerator to the coldest setting.

Move car inside garage and secure door.

Keep trees and shrubs well trimmed.

Bring in all outdoor furniture and anything else not secured.

Fill car with fuel.

Plan an evacuation plan and where to meet family.

Figure 3.9 How planning can help reduce hazard risk

Now test yourself

1 Define a) monitoring, b) prediction, c) protection, d) planning.
2 How does a satellite help to predict tropical storms?
3 What is a forecast cone?
4 Why would it be helpful when preparing for a tropical storm to a) store loose objects, and b) have fuel in vehicles?

> **Exam tip**
>
> Look at exam practice question 1. Make sure you answer all of the question by stating which effects are reduced.

Exam practice

1 Explain how protection against tropical storms can reduce the effects of tropical storms. (6 marks)
2 Use a case study of a tropical storm to describe the primary and secondary effects. (9 marks)

> **Exam tip**
>
> Use facts and figures that you have learnt in case-study questions to prove you are writing specifically about this place and not any other.

3.4 Weather hazards in the UK

What are weather hazards in the UK?

Extreme weather is when the weather is especially severe or out of season, and is clearly different to the usual weather pattern. Most parts of the UK are at risk from one or more types of extreme weather. Different air masses crossing the UK bring a variety of weather.

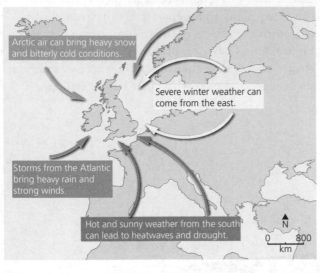

Arctic air can bring heavy snow and bitterly cold conditions.

Severe winter weather can come from the east.

Storms from the Atlantic bring heavy rain and strong winds.

Hot and sunny weather from the south can lead to heatwaves and drought.

N

0   800
km

Figure 3.10 Map of UK air masses

Storm events

Depressions (low-pressure systems) bring heavy rain and strong winds to the UK, as happened in autumn 2013. Possible impacts include:

- flood and wind damage to business and properties
- trees uprooted, causing further damage
- power supplies down
- disruption to transport
- death.

Flooding

Floods are often caused by heavy rainfall or storm waves. Torrential rainstorms and thunderstorms can cause flash flooding. Prolonged rainfall also leads to flooding, for example as in Boscastle, 2004. Possible impacts include:

- crops ruined
- damage to homes, businesses and possessions
- transport disrupted or even washed away
- death by drowning
- recovery is often expensive
- landslides.

Droughts and heatwaves

Droughts and heatwaves are typically long periods with little or no rainfall. In the UK a drought is defined as fifteen or more consecutive days with less than 0.2 millimetres of rain on any one day, such as in summer 2003. Possible impacts include:

- crop production fails and wildlife is affected
- reservoirs run low, reducing water supplies
- hosepipe bans enforced
- elderly vulnerable to heat exhaustion – possible death
- roads can melt and railway lines can buckle.

However, the tourism industry may benefit.

Extremes of cold weather

Cold conditions take over if the usual depressions (low-pressure systems) are not passing over the UK, as happened during winter 2014–15. Possible impacts include:

- crops fail and cattle may not survive at −10°C
- roads, railways and airlines shut
- increased injuries caused by falling in snow and ice
- businesses and schools shut.

Now test yourself

1 Describe why the UK experiences a variety of weather.
2 What is extreme weather?
3 Name four types of extreme weather that the UK experiences.
4 How could extreme weather affect a) agriculture b) transport c) people's health?
5 Why might extreme weather be of benefit to the UK?

Revision activity

Practise sketching a simple map of the UK and label it with where different types of weather come from.

What have been the recent extreme weather events in the UK?

Both Cumbria and the Somerset Levels are examples in the UK which experienced an extreme weather event leading to widespread flooding. You are required to study **one** example of an extreme weather event in the UK.

	Somerset Levels, UK	Cumbria, UK
When	December 2013 to February 2014	17–20 November 2009
Causes	• Several depressions moving across the Atlantic Ocean caused weeks of wet weather. • Saturated soil. • Wettest January on record. • High tides and storm surges came up the rivers from the Bristol Channel. • Reduced river capacity (lots of sediment) due to not being dredged for over twenty years.	• A month's worth of average rainfall had fallen by 17th November. • Saturated soil. • Steep slopes of the Lake District. • Deep Atlantic depression moving north-eastwards over Scotland and northern England. • Heaviest rainfall on record.
Social impacts	• More than 600 homes flooded. • Sixteen farms evacuated. • Temporary accommodation for residents needed for several months. • Some villages cut off. • Power supplies disrupted.	• More than 1,500 homes flooded. • Police officer killed when a bridge in Workington collapsed. • Many injured. • Health risk as river water contaminated with sewage.
Economic impacts	• Over 14,000 hectares of agricultural land flooded for weeks. • Over 1,000 livestock evacuated. • Roads cut off. • Railway line closed • £10 million damage estimated.	• Businesses closed, which didn't reopen until long afterwards. • Six important regional bridges destroyed. • £100 million damages.
Environmental impacts	• Contaminated river water with sewage, oils and chemicals. • Large volume of debris deposited across land. • Stagnant water had to be reoxygenated then pumped back into river.	• Landslides triggered. • Contaminated river water. • Hundreds of trees carried away, damaging local ecosystems and habitats.
Management strategies to reduce risk	• River banks raised and strengthened. • Somerset County Council pledged £20 million on a Flood Action Plan. • Rivers Tone and Parratt were dredged in March 2014. • Road levels raised. • Flood defences for communities at risk. • Pumping stations built. • Potential tidal barrage at Bridgwater by 2024.	• Central government and local community spent £4.5 million on new flood defences. • Mobile wall built, which rises when needed but disappears from view to maintain views for tourist industry. • Environment Agency (EA) provides residents with improved flood warning information. • The EA sends flood warning messages directly to smartphones of Cockermouth residents. • Adverts placed in national newspapers to publicise Cumbria was back open for business.

Now test yourself

1 State three factors that caused flooding.
2 Name two social, economic or environmental impacts.
3 Who was involved in reducing the risk of future floods?

Revision activity

Write a series of quiz questions about your case study. Test yourself over and over until you consistently remember the answers correctly.

Is the UK's weather getting more extreme?

Extreme weather is not new to the UK. There are many examples of extreme weather in the past. However, the frequency of extreme weather in the UK is increasing. Since the 1980s, UK temperatures have increased by about 1°C and winter rainfall has increased. There have been more weather records broken recently than ever before.

Extreme weather records

Temperature	Rainfall
● December 2010 coldest on record for 100 years. Warmest April was 2011. ● Highest temperature (38.5°C) was 10 August 2003. ● Lowest temperature (–27°C) was in Scotland in 1995.	● Highest two-day record of rainfall (405 millimetres) was in 2015. ● Highest three- and four-day rainfall records were both in 2009. ● Highest monthly total of rainfall (1,396.4 millimetres) was in 2015. ● Serious flooding has become more frequent in the winter, for example in Cumbria 2009 and the Somerset Levels 2013–14.

What are the predictions for future UK weather?

● Precipitation is expected to become even more seasonal.
● Some rivers are expected to flood more frequently in winter.
● Air temperature is expected to increase, causing more drought.

Is climate change responsible?

Climate change cannot be responsible for individual extreme weather events. Yet scientists suggest that the increasing frequency of extreme weather events can be blamed on climate change. Evidence suggests that climate change is warming the planet. The Atlantic Ocean is increasing in temperature. This can explain the UK's changing rainfall pattern. Rain-bearing depressions will gain more energy and moisture, due to the warmer ocean. The frequency of rain-bearing storms has increased in line with climate change predictions since the 1980s.

Research also shows that the Atlantic Ocean was relatively warm between 1931 and 1960, before cooling from 1961 and 1990, that and then warming again. The increase in ocean temperature may simply be a long-term cycle or due to the position of the jet stream and not due to climate change.

Now test yourself

1 What has happened to the frequency of extreme weather events in the UK?
2 By how much have temperatures increased since the 1980s?
3 Name two examples of UK weather records that have been broken since 2000.
4 Describe how a) precipitation, b) flooding, and c) air temperature are expected to change in the future.
5 Give one reason why climate change is not responsible for the UK's extreme weather.

Exam practice

1 Give one piece of evidence that suggests the UK's weather is becoming more extreme. (1 mark)
2 Suggest why the UK's extreme weather events might be increasing. (4 marks)
3 Discuss to what extent climate change is responsible for extreme weather in the UK. (6 marks)

4 Climate change

4.1 Causes and effects of climate change

What are the possible causes?

Evidence of climate change occurring before humans existed means climate change must be natural. However, natural causes cannot account for the unprecedented temperature increase since the 1970s. A thicker layer of greenhouse gases (carbon dioxide 77 per cent, methane 14 per cent, nitrous oxide 8 per cent and CFCs 1 per cent) caused by human activity means less of the Sun's energy is able to escape the Earth's atmosphere, so the temperature increases.

> **Revision activity**
>
> Draw a pie chart to show the proportions of each of the greenhouse gases.

What is the evidence for climate change?

Since 1914 the Met Office has had reliable climate-change data using weather stations, satellites, weather balloons, radar and ocean buoys. Evidence includes:
- an increase in average surface air temperature by 1°C over the last 100 years
- the warmest ocean temperatures since 1850
- a 19-centimetre rise in sea levels since 1900.

Natural recorders, such as tree rings, ice cores (spanning 800,000 years) and ocean sediments (spanning beyond the **Quaternary period**), help estimate climate. Oxygen, carbon dioxide and methane in ice cores and ocean sediments can help estimate past temperature by comparing it to present levels. Organisms and plankton in ocean sediments reveal water temperatures, oxygen levels and nutrients, and these can indicate climate change.

The Quaternary period has over twenty cycles of cold glacial periods (lasting about 100,000 years) and warmer interglacial periods (lasting about 10,000 years). The current interglacial period has lasted 15,000 years.

Natural factors

Orbital changes

The Sun's energy on the Earth's surface changes as the Earth's orbit is elliptical, its axis is tilted on an angle and the Earth isn't spherical.

Solar output

Sunspots increase from a minimum to maximum every 11 years. Fewer sunspots were observed during the coldest period ('Little Ice Age' in 1645–1715). However, solar output has barely changed in the last 50 years.

Causes of climate change

Human factors

Fossil fuels

Burning fossil fuels releases carbon dioxide. Accounts for 50% of greenhouse gases.

Agriculture

Accounts for 20% of greenhouse gases. Larger populations and demand for meat and rice will increase its contribution.

Volcanic activity

Volcanic aerosols reflect sunlight away, reducing global temperatures temporarily.

Deforestation

Logging and clearing land for agriculture/roads increase the amount of CO_2 in the atmosphere, as less photosynthesis occurs.

Figure 4.1 Causes of climate change

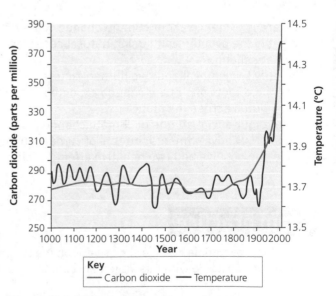

Figure 4.2 Global temperature change

What are the effects of climate change?

REVISED

The effects of climate change are not certain. They are likely to be unevenly distributed across the world and will depend on the human and physical circumstances of the location. For example, low-lying coastal countries will be more vulnerable to flooding and poorer countries have less ability to invest in prediction and protection strategies.

Social effects (effects on people)

- Increased risk of diseases such as skin cancers and heat stroke as temperatures increase.
- Winter-related deaths decrease with milder winters.
- Crop yields affected; maize will decrease by up to 12 per cent in South America, yet will increase in northern Europe and require more irrigation.
- Less ice in the Arctic Ocean increases shipping and extraction of gas and oil reserves.
- Drought reduces food and water supplies in sub-Saharan Africa. Water scarcity in the south and southeast of the UK.
- Flood risk increases repair and insurance costs. Seventy per cent of Asia at increased risk of flooding, causing migration and overcrowding in low-risk areas.
- Declining fishing in the Lower Mekong delta would affect 40 million, due to changing water quality because of reduced water flow and sea-level rise.
- Extreme weather increases investment in prediction and protection.
- Skiing industry may decline in Alps as less snow.

Environmental effects

- Increased drought in areas such as Mediterranean region.
- Lower rainfall causes food shortages for orang-utans in Borneo and Indonesia.
- Sea-level rise increases flooding and coastal **erosion,** so defences are under increasing strain.
- Ice melts, so wildlife declines, such as Adélie penguins on the Antarctic Peninsula and polar bears and seals in the Arctic.
- Warmer rivers affect marine wildlife; for example, the food supply will decrease for the Ganges river dolphin.
- Increase in forest growth in northern Europe.
- Forests in North America may experience more pests, disease and forest fires.
- Coral bleaching, and decline in biodiversity such as at the Great Barrier Reef.

Now test yourself

TESTED

1 Where is evidence about climate change collected from?
2 What is the problem with using natural recorders as evidence for climate change?
3 Identify two natural and two human factors causing climate change.
4 Are the effects of climate change expected to be evenly or unevenly distributed across the globe?
5 Name four social effects of climate change.
6 Name four environmental effects of climate change.
7 Give one positive and two negative effects of climate change.

Exam practice

'The effects of climate change are greater on the environment than on people.' Do you agree with this statement? Justify your decision. (9 marks)

ONLINE

4.2 Managing climate change

How can climate change be mitigated?

REVISED

There are various ways **mitigation** can help reduce climate change:

1 Alternative energy production
Alternative energy production (such as wind, solar, geothermal, wave and tidal, and biomass) reduces greenhouse gases compared to burning fossil fuels (coal, oil and gas). They will last longer. However, despite becoming cheaper and more competitive, they are expensive and cannot be relied upon to generate electricity if, for example, there is no wind, sun or waves.

2 Carbon capture
Carbon capture takes carbon dioxide (CO_2) from emission sources and safely stores it underground. An impermeable 'cap rock' prevents it escaping. It can capture up to 90 per cent of CO_2 and provide 10–50 per cent of the world's total carbon mitigation until 2100. However, it is expensive, it is unclear if CO_2 would remain captured long term, and it doesn't promote renewable energy.

Mitigation

3 Planting trees
Planting trees helps to remove CO_2 from the atmosphere through photosynthesis. It could increase forest carbon storage by 28 per cent. Oxygen is produced during photosynthesis, and trees provide habitats. However, land may be limited and biodiversity is reduced if only one tree species is planted.

4 International agreements
International agreements encourage countries to take responsibility for reducing CO_2 emissions. Targets are more likely to be met if legally binding (Paris 2015 agreement). Financial support may be provided for LICs. However, some countries are considered more responsible, it is hard to agree targets that go far enough, and they may not be achieved.

Figure 4.3 Mitigation can help climate change

How can we adapt to climate change?

REVISED

There are various ways **adaptation** can help reduce climate change:

1 Changes in agricultural systems
Changes in agricultural systems are required to deal with changing rainfall and temperature patterns, weather becoming more extreme and the changing distribution of pests and diseases. Production may need to move location to suit climates, irrigation may be necessary and changes to crops and varieties may be required. These adaptations are most difficult for poorer farmers, who are most likely to be affected.

2 Managing water supplies
Managing water supplies ensures populations can face the challenge of changing rainfall patterns. In London, this involves reducing demand (such as installing water-efficient devices) and increasing supply (such as opening a desalination plant). In addition to water supplies being under strain, security may be threatened in areas of deficit, especially where there is less political stability.

Adaptation

3 Reducing risk
Reducing risk from rising sea levels could involve constructing defences (such as the Thames Barrier or restoring mangrove forests), raising properties on stilts or relocating people at risk. There are economic, social and environmental implications of these strategies.

Figure 4.4 Adaptation can help climate change

TESTED

Now test yourself

1 Define mitigation.
2 Define adaptation.
3 Name two methods of mitigation and adaptation.
4 Describe how water supply can be managed.

Revision activity

Try making up a sentence to remember the four mitigation strategies and the three adaptation strategies.

Exam practice

1 Compare and contrast mitigation and adaptation.
(2 marks)
2 Outline one possible method of climate change mitigation. (2 marks)

ONLINE

5 Ecosystems

5.1 Ecosystems

What is an ecosystem?

REVISED

An **ecosystem** is a natural system that is made up of plants, animals and the environment in which they live. The various components of an ecosystem – climate, water, soil, plants, animals and people – are closely interlinked and depend on one another for survival. If one component changes, there will be knock-on effects within the ecosystem.

It is possible to identify two types of component in an ecosystem:
- **Biotic** – such as plants and animals, bacteria and fungi.
- **Abiotic** – such as climate, water and soils.

A small-scale ecosystem: a freshwater pond

Figure 5.1 is an example of a small-scale ecosystem in the UK. Notice that there are examples of **producers**, **consumers** and **decomposers**. All three are vital components in a healthy and sustainable ecosystem.

> **Now test yourself**
>
> Identify the producers, consumers and decomposers in Figure 5.1.
>
> TESTED

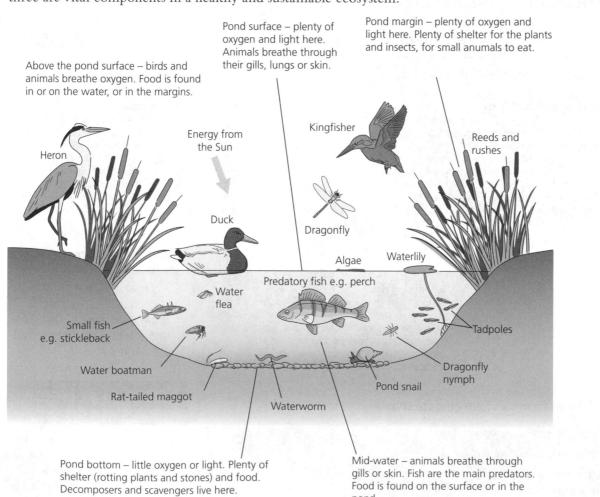

Pond surface – plenty of oxygen and light here. Animals breathe through their gills, lungs or skin.

Pond margin – plenty of oxygen and light here. Plenty of shelter for the plants and insects, for small anumals to eat.

Above the pond surface – birds and animals breathe oxygen. Food is found in or on the water, or in the margins.

Energy from the Sun

Kingfisher

Reeds and rushes

Heron

Duck

Dragonfly

Algae

Waterlily

Predatory fish e.g. perch

Water flea

Small fish e.g. stickleback

Tadpoles

Water boatman

Dragonfly nymph

Rat-tailed maggot

Pond snail

Waterworm

Pond bottom – little oxygen or light. Plenty of shelter (rotting plants and stones) and food. Decomposers and scavengers live here.

Mid-water – animals breathe through gills or skin. Fish are the main predators. Food is found on the surface or in the pond.

Figure 5.1 A freshwater pond ecosystem

Food chains and food webs

The links between the biotic components in an ecosystem can be shown by two types of flow diagram:

- **Food chain** (Figure 5.2) – this shows the direct links between different organisms that rely on each other as a source of food (essentially who is eating whom!).
- **Food web** (Figure 5.3) – this shows the complex hierarchy of plants and animals that rely on each other as a source of food within an ecosystem (rather like a menu!).

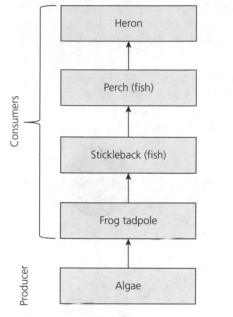

Figure 5.2 A freshwater pond food chain

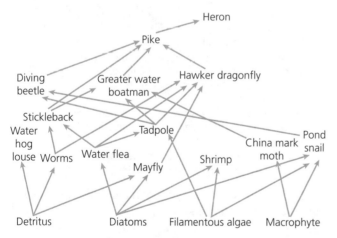

Figure 5.3 A freshwater pond food web

Food chains and food webs are extremely important in understanding the possible impacts of change within an ecosystem. The sudden removal of one species or massive growth of another can have huge impacts on other components within the ecosystem.

What is nutrient cycling? REVISED

Nutrients are foods that are used by plants and animals as they grow. Nutrients are derived from two main sources:

- Rainwater, washing chemicals out of the atmosphere.
- Weathering of rocks, releasing chemicals into the soil.

In a typical **nutrient cycle**, there are three main nutrient stores, and several flows responsible for transferring nutrients between the stores. There are also flows to and from the abiotic components, such as the rock.

Example

College Lake, Buckinghamshire

College Lake is a protected nature reserve managed by the Berks, Bucks and Oxon Wildlife Trust. It is a former chalk quarry that has been restored to create a variety of habitats, including a lake:

- The open water and marshland provides important roosting and breeding areas for native and seasonal birds.
- The wigeon, a bird native to Iceland, Scandinavia and Russia, spends winters in the UK feeding on aquatic plants (producers) near the margins of the lake and its islands.
- In the summer the many small islands in the lake are home to lapwing and redshank. Their chicks feed in the muddy margins, which are full of colourful flowers and insects.
- Alder grows well at the lake margins, providing food for several species of caterpillars, seed for birds and nectar and pollen for bees (consumers). Moorhens nest around the roots and vegetation on the lake margins.
- In addition to the birds, flowers and insects, there are also fish and amphibians (consumers).

As you can see, there are important interrelationships at College Lake. Additionally, the Trust manages the reserve to protect the wildlife while encouraging people to visit. Visitors play an important role in seed dispersal, as seeds become attached to their clothing while they walk around the lake and are then dropped elsewhere.

Revision activity

Draw a simple diagram to show nutrient recycling, using different colours to represent the different components.

Now test yourself TESTED

Identify a food chain in Figure 5.3.

Now test yourself TESTED

Identify producers and consumers in College Lake.

Example

Epping Forest, Essex

Epping Forest is an ancient deciduous woodland to the northeast of London. It has:
- several native tree species including oak, elm, ash and beech
- a lower shrub layer of grasses, brambles and bracken that form the main producers
- many insects, mammals, amphibians and birds, which are the consumers
- over 700 species of fungi, which are important decomposers in the ecosystem.

The interrelationships between producers, consumers and decomposers, together with the abiotic factors, can be illustrated by the annual life cycle of the deciduous trees.

Season	Ecosystem interrelationships: deciduous trees
Spring	Flowering bulbs such as bluebells make use of the sunlight penetrating through branches. The stored nutrients are used by the growing plants (producers) to produce fruit, berries and nuts that will feed consumers.
Summer	The broad tree leaves grow quickly in the spring. With a large surface area, they maximise the Sun's energy to photosynthesise.
Autumn	To conserve energy and moisture, the trees shed their leaves. This is a direct response to the UK's climate, as the temperature and hours of sunlight decrease.
Winter	Bacteria and fungi decompose the leaf litter, releasing the nutrients into the soil.

Revision activity

Draw a simple diagram to show nutrient recycling, using different colours to represent the different components.

People also have an important role to play:
- Epping Forest has been managed for centuries, initially as a hunting forest for royalty, then for its timber, and now for recreation and conservation.
- Many trees have been coppiced (cut down to ground level) or pollarded (cut down to shoulder level) to encourage new, straight growth for timber.
- Visitors pick fruit and berries, helping to disperse seeds.

The balance and interdependence between the components of the deciduous woodland ecosystem can be demonstrated by the nutrient recycling that takes place (Figure 5.4).

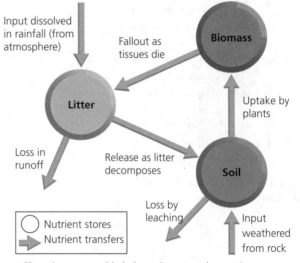

- There is a reasonable balance between the nutrient stores, with biomass the largest, due to rapid vegetation growth in favourable climatic conditions and fertile soils.
- There are moderate transfers of nutrients between the stores, reflecting the vigorous cycle of growth and rapid leaf decomposition.
- Leaching (nutrients dissolved and carried away by water) is moderate, reflecting high rainfall.

Figure 5.4 Nutrient recycling in a deciduous forest such as Epping Forest

What is the impact of change on an ecosystem?

A healthy and sustainable ecosystem can take many hundreds of years to develop. Its success depends on the many complex interrelationships that exist within it. While all ecosystems evolve slowly and adapt to change, a sudden change can have disastrous impacts, causing it to become unbalanced.

Natural change	Human-induced change
<ul><li>Extreme weather events, such as flooding and drought.</li><li>Fire caused by lightning strikes.</li><li>Climate change and global warming.</li><li>Spread of invasive species or introduction of alien species.</li></ul>	<ul><li>Land use change, such as deforestation and hedgerow removal.</li><li>Alteration to water and soils, such as land drainage or adding fertilisers.</li><li>Hunting or trapping of animals or birds.</li><li>Introduction of alien species.</li></ul>

How can changes affect the pond ecosystem?

The key component of a pond ecosystem is water! If the quantity or quality of water changes, this can have a massive impact on the ecosystem.

- Climate change, and extreme weather events such as heatwaves and droughts, can dramatically affect water levels. Long periods of very low water levels will cause stress in organisms, and some individuals may die. If a pond dries out, the entire ecosystem will be at risk.
- Fertilisers on farmland can result in an influx of nutrients into a pond. This leads to the rapid growth of algae, which effectively chokes the pond, depriving other organisms of oxygen. This process is called eutrophication.
- Introduction of alien species may happen naturally or through human activities. For example, if alien species, such as predatory fish like pike or perch, are introduced, this could reduce the food supply (small fish and frogs) for organisms further up the food chain, for example herons.

> **Revision activity**
>
> Look back to the College Lake example on page 21. Use a flow diagram to show how the removal of alders could affect the pond ecosystem.

How can changes affect the deciduous forest ecosystem?

- Chalara is a disease affecting many ash trees. It is caused by a fungus and results in tree loss and bark damage. These trees are also being affected by a borer-beetle. The lack of ash trees in a woodland will affect the ecosystem's primary production and may reduce food for consumers.
- In 1987 strong winds blew down over 15 million trees in southern England. This caused forest ecosystems to become unbalanced, with consumer species declining. However, secondary growth of trees has restored the balance, with consumer numbers increasing once more.
- Controlled hunting (culling) for species such as deer (consumer) can be important in maintaining balance within an ecosystem. If deer numbers are allowed to escalate, this can reduce the food supply (producers) for other consumers.

Figure 5.5 Algal growth in a pond

What are global ecosystems?

Ecosystems exist at all scales, from small-scale ponds and hedgerows, through to large-scale **global ecosystems** or biomes, such as tropical rainforests and hot deserts.

Figure 5.6 shows the distribution of the global ecosystems.

The primary factor affecting the pattern of global ecosystems is climate. This is why most global ecosystems form latitudinal belts across the world (look at Figure 3.2, page 8, to see the links with global atmospheric circulation).

Variations in this pattern reflect other factors, such as warm and cold ocean currents, the distribution of land and sea, and the pattern of surface winds. These factors result in small-scale variations in weather and climate, which impact on ecosystems. For example, the UK experiences a temperate maritime (warm and wet) climate, due to the prevailing south-west winds and the presence of the North Atlantic Drift, which brings warm waters from the Caribbean to the west coast of the UK.

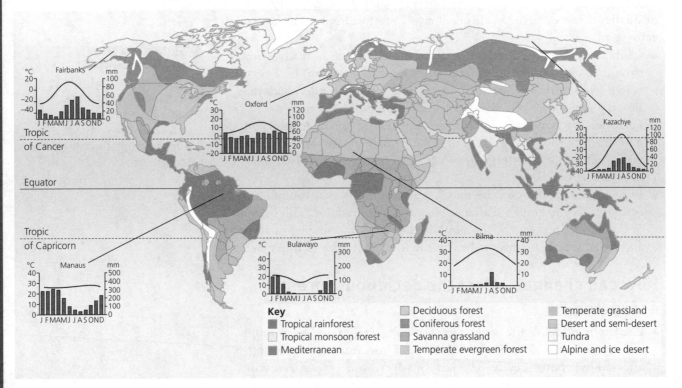

Key
- Tropical rainforest
- Tropical monsoon forest
- Mediterranean
- Deciduous forest
- Coniferous forest
- Savanna grassland
- Temperate evergreen forest
- Temperate grassland
- Desert and semi-desert
- Tundra
- Alpine and ice desert

Figure 5.6 The distribution of global ecosystems. Remember that the different zones merge into one another, rather than having sharp boundaries as the map suggests

Global ecosystem	Location	Characteristics
Coniferous forest	Roughly 60 degrees north	Cold and dark winters but quite warm summers; not as extreme as **tundra**. Coniferous trees are cone-bearing and many are evergreen so that they can photosynthesise immediately after the winter.
Deciduous forest	Roughly 50 degrees north, the natural ecosystem for the UK and much of western Europe	Deciduous trees shed their leaves in winter to retain moisture. The climate here is more moderate, with mostly mild and moist conditions and few extremes of temperature.
Desert	Roughly 30 degrees north and south of the Equator close to the Tropics of Cancer and Capricorn	Sinking air in these latitudes suppresses rain formation, leading to arid conditions. Hot in the daytime but cooler at night, due to the lack of cloud cover, allowing heat to escape. Plants and animals have to be very well adapted to cope with these harsh conditions.
Mediterranean	Roughly 40–45 degrees north of the Equator, centred on the Mediterranean. Isolated pockets in South Africa and Western Australia	Hot and dry summers, wet and mild winters typical of countries such as Greece and southern Spain. Vegetation includes citrus fruit trees, oaks and olives.
Polar/tundra	Arctic and Antarctic (polar) and high latitudes, such as Canada and Siberia (tundra)	Extremely cold throughout the year in polar regions; cold winters but quite warm brief summers in tundra regions. Limited precipitation. Tundra is very fragile and easily damaged by human activities, e.g. oil exploitation.
Temperate grassland	Roughly 30–40 degrees north and south of the Equator in continental interiors	Continental conditions result in hot summers and cold winters with relatively low rainfall. Ideal conditions for grasses and grazing animals.
Tropical rainforest	Close to the Equator, widespread across Asia, Africa and South America	Concentrated energy from the Sun heats the moist air, which rises to produce heavy rainfall. This, combined with high temperatures, means the conditions are ideal for plant growth. Rainforests cover 6% of the Earth's land surface. Over 50% of the world's plants and animals live in this ecosystem.
Tropical grassland (savanna)	Between 15–30 degrees north and south of the Equator	The tropical climate in these low latitudes is characterised by distinct wet and dry seasons. Fires are common in the dry season, usually ignited by lightning strikes. Herds graze these areas, along with predators such as lions.

Now test yourself

 TESTED

1 How can extreme weather events lead to change in a freshwater pond ecosystem?
2 Apart from the climate, what other factors affect the pattern of global ecosystems?

Exam practice

1 What is a decomposer? (2 marks)
2 Describe a food chain for a chosen small-scale ecosystem in the UK. (4 marks)
3 Study Figure 5.6. Compare the distribution of the coniferous forest and the tropical rainforest global ecosystems. (4 marks)
4 To what extent is change a natural part of the development of an ecosystem? (6 marks)

ONLINE

6 Tropical rainforests

6.1 Tropical rainforests

Where are tropical rainforests found?

REVISED

Tropical rainforests cover about 2 per cent of the Earth's surface yet they are home to over half of the world's plants and animals. They are found in a broad belt close to the Equator from South America in the west, through West Africa to Southeast Asia and Australia in the east (Figure 6.1).

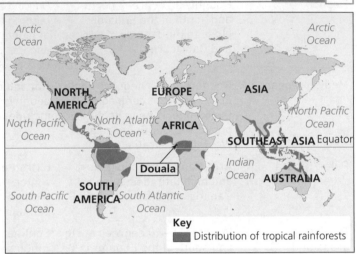

Figure 6.1 The global distribution of tropical rainforests

What are the physical characteristics of tropical rainforests?

REVISED

Climate

Tropical rainforests thrive in the equatorial climate, experiencing high temperatures (about 27°C) throughout the year and high rainfall (over 2,000 millimetres per year). This climate creates ideal growing conditions and accounts for the lush vegetation growth. Figure 6.2 is a climate graph for Douala in Cameroon. Its location is shown on Figure 6.1.

Water

Most tropical rainforests experience a distinct wet season, with high rainfall totals lasting for several months. You can see this in Figure 6.2. During this time, there is excess water on the ground, swelling local rivers and sometimes causing flooding. Water will soak into the soil, dissolving and transporting away nutrients – a process called leaching.

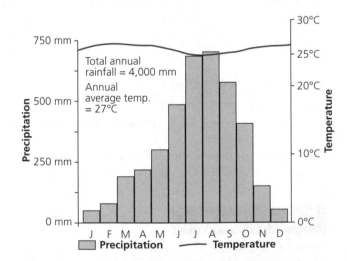

Figure 6.2 Climate graph for Douala, Cameroon

Now test yourself

Identify the wet season in Douala using Figure 6.2. Which month has the highest rainfall, and approximately how much is it?

TESTED

Soils

Rainforest soils are not very fertile. Nutrients are concentrated in the upper topsoil and are very quickly taken up by the plants as they grow. In response to this, trees and plants have shallow roots to maximise their use of these vital nutrients. Intense leaching removes nutrients from the topsoil and redeposits them further down. This accounts for the red-coloured iron-rich soils (called latosols) that characterise tropical rainforest environments.

Plants and animals

Rainforests have the highest level of **biodiversity** in the world. There are thousands of species of plants and animals, and scientists believe that many have yet to be discovered! The reason for this is the huge range of habitats available.
- Birds live in the canopy (branches and leaves) high above the forest floor, feeding on seeds and nectar from flowering plants.
- Mammals, such as monkeys and sloths, are well adapted to life in the trees.
- Snakes use the trunks of trees as vertical highways from the forest floor to the canopy above.
- Animals such as deer and rodents browse on vegetation on the forest floor.

People

Traditional tribes living in rainforests live in harmony with the natural environment, hunting and gathering only what they need to survive. This is a **sustainable** system. However, increasingly people are exploiting rainforests for commercial gain, chopping down trees for timber or to make way for ranching or commercial plantations. This is extremely harmful to the rainforest ecosystem and reduces biodiversity as habitats are destroyed.

How is the tropical rainforest interdependent?

The tropical rainforest ecosystem is a **fragile environment** – it is easily damaged by people's activities and can take a long time to recover. Its survival depends on a close interdependence between its component parts. If one component changes – for example, trees are removed through deforestation – harmful knock-on effects will threaten the survival of the entire ecosystem.

Nutrient recycling

Figure 6.3 shows nutrient recycling in a tropical rainforest ecosystem. It shows very clearly the interdependence between the physical characteristics of tropical rainforests.
- The vast majority of nutrients are stored in the biomass (mostly the trees).
- The soil contains few nutrients, as any nutrients released by decomposers are quickly absorbed by the trees and plants or leached into the soil by the heavy rainfall.
- There are few nutrients in the litter store, as decomposers, thriving in the warm and wet conditions, quickly break down dead leaves and branches.
- There is a rapid transfer of nutrients indicated by the thick arrows. This is largely due to the climatic conditions; for example, warm and wet conditions promote chemical weathering.

Now test yourself

Why do plants and trees in the rainforest tend to have shallow roots?

TESTED

Revision activity

Draw a spider diagram and use labelled arrows to show the interdependence of climate, water, soils, plants, animals and people in a tropical rainforest.

REVISED

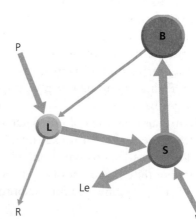

Figure 6.3 A tropical rainforest nutrient cycle

What are plant and animal adaptations?

Plant adaptations

Tropical rainforests are characterised by a layered, or stratified, structure (Figure 6.4).

- The vegetation, and animals, are well adapted to the climatic conditions experienced within each layer.
- Access to sunshine is a key factor promoting strong vertical growth, as shown by the tallest emergent trees 'punching' their way through the main canopy to exploit maximum sunlight.
- The many and varied adaptations show the interdependence between plants, climate and soils.

> **Exam tip**
>
> When explaining vegetation adaptations, make sure you link them to physical conditions, such as climate or soil. When describing plant or animal adaptations, make sure you say why the adaptation has developed.

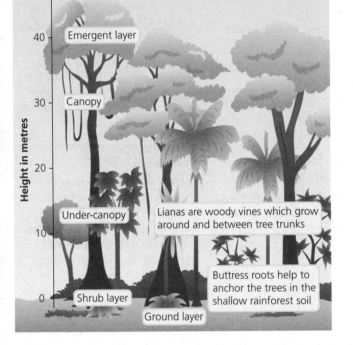

Figure 6.4 Structure of the rainforest

Labels in figure:
- Emergent layer
- Canopy
- Under-canopy
- Shrub layer
- Ground layer
- Lianas are woody vines which grow around and between tree trunks
- Buttress roots help to anchor the trees in the shallow rainforest soil
- Height in metres (0, 10, 20, 30, 40)

> **Now test yourself**
>
> Identify and explain two vegetation adaptations shown in Figure 6.4.
>
> TESTED

Animal adaptations

While there is plenty of food and water throughout the year, animals show special adaptations to living in tropical rainforests.

- Due to the intense competition for food, animals have become adapted to live off specific plants or animals that few others eat. For example, toucans and parrots have strong beaks for cracking hard nuts which other birds cannot break.
- Bats thrive on the fruits growing in the canopy, which they can reach by flying. They also help to disperse seeds, which pass through their digestive systems.
- Animals may use colour to act as camouflage or to warn predators to leave them alone; for example, brightly coloured poisonous frogs.
- Three-toed sloths have long claws, enabling them to climb trees, where they live most of the time, away from forest-floor predators. Algae grow on their fur, helping to camouflage them!

Figure 6.5 Sloths have adapted to living in forests

> **Now test yourself**
>
> Describe the special adaptations of toucans, poisonous frogs and three-toed sloths.
>
> TESTED

What are biodiversity issues?

How biodiverse are tropical rainforests?

No one knows exactly how many species of plants and animals live in tropical rainforests – but we do know that, per area, they are the most biodiverse ecosystems in the world. For example:

- Indonesian rainforests are estimated to have over 30,000 species of plants and over 1,600 species of birds.
- Rainforests contain 170,000 of the world's 250,000 known plant species.
- A recent survey found 487 separate tree species in a single hectare in Brazil; this compares with about 700 tree species throughout the whole of North America.

Why are there high levels of biodiversity in tropical rainforests?

There are several reasons why there is a high level of biodiversity in tropical rainforests:

- The wet and warm climate encourages a wide range of plants and trees to grow. These provide many different natural habitats for animals.
- The rapid recycling of nutrients speeds up plant growth (producers) and provides plentiful food for consumers.
- Many parts of tropical rainforests are untouched by people, enabling a range of plants and animals to thrive.

What are the threats to biodiversity in tropical rainforests?

With the exception of natural events such as lightning strikes (fires), floods and disease, the greatest threats come from people, as seen in Figure 6.6.

What are the main issues associated with biodiversity decline?

One of the most important components of an ecosystem is a keystone species: a species that has multiple connections with other species. As it is a multiple food source, there could be serious knock-on effects if its numbers decline.

For example, in Borneo, fig trees are pollinated by fig wasps and the seeds are dispersed by orang-utans. Orang-utan numbers have declined hugely in recent years, due to habitat destruction and hunting. In the future, this will have an effect on fig tree propagation, which in turn will affect all species dependent on the fig tree.

Other issues include the following:

- Indigenous tribes being unable to survive in rainforests and having to abandon their traditional lifestyle.
- Plant and animal species may become extinct – some even before they have been discovered.
- Important medical plants may become extinct.

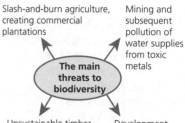

Figure 6.6 The main threats to biodiversity

6.2 What is deforestation?

How is the rate of deforestation changing?

Tropical rainforests are found in over 60 countries in the world. The greatest concentration in terms of biomass is in Brazil, followed by the Democratic Republic of Congo and Indonesia (Figure 6.7). In terms of hectares, Brazil has about 480,000 hectares of rainforest, compared to about 125,000 hectares in Congo and about 80,000 hectares in Indonesia.

As Figure 6.8 shows, the greatest loss of rainforest has been recorded in Brazil and Indonesia. Since 1970 about 20 per cent of the Amazon rainforest has been cleared, roughly three times the size of the UK.

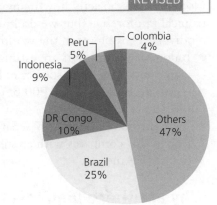

Figure 6.7 Tropical rainforest biomass by country

Rainforest countries with most forest loss, 2000–2012 (sq km)

Country	sq km
Brazil	360,227
Indonesia	157,850
DR Congo	58,963
Malaysia	47,278
Bolivia	29,867
Colombia	25,193
Peru	15,288
Myanmar	14,958
Cote d'Ivoire	14,889
Madagascar	14,659
Venezuela	12,958
Cambodia	12,595
Vietnam	12,289
Laos	12,084

Figure 6.8 Deforestation in selected countries

In recent decades, the rate of deforestation has changed (Figure 6.9).

- Despite the huge losses in Brazil, the rate of deforestation has started to decline. Over half of the remaining rainforest in Brazil is now protected from deforestation.
- In contrast, the rates of deforestation have increased significantly in Indonesia and in the Peruvian Amazon (Figure 6.9).
- Malaysia has one of the fastest-growing rates of deforestation – between 2000 and 2013 an area the size of Denmark was deforested.
- Even where the rates of deforestation have decreased, the practice is still continuing, with an estimated 31 million hectares being felled each year.

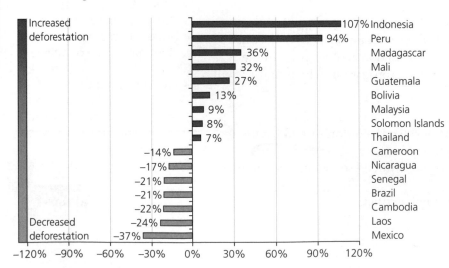

Figure 6.9 Change in the annual deforestation rate (2000–10)

Now test yourself

1 Why do you think rates of deforestation are increasing in some parts of the world but decreasing in others?
2 Why is it so important that the rate of deforestation in Brazil has started to fall?
3 Study Figure 6.8.
   a) Calculate the amount of rainforest lost in South America.
   b) Calculate the forest loss in Indonesia as a percentage of that lost in Brazil.
4 Study Figure 6.9. Calculate the difference in the rate of deforestation between Indonesia and Mexico.

Cause of deforestation	Description
Subsistence farming and commercial farming	Indigenous rainforest tribes practise **subsistence farming**. However, increasingly land is cleared to make way for **commercial farming** – crop plantations and cattle grazing. ● In Brazil, cattle ranching accounts for 80% of deforestation. In Brazil, soybeans, palm oil and sugar cane (for biofuel) are the major crops. ● Malaysia is the largest exporter of palm oil in the world. Huge areas of rainforest have been converted since the 1970s. Tax incentives for plantation owners encourage further development.
Logging	Trees such as mahogany and teak are highly valued for furniture and other uses. Smaller trees are used for fuel, pulped or made into charcoal. Malaysia is one of the world's largest exporters of tropical wood. Much of it involves clear felling, causing total habitat destruction. Around 80% of deforestation in Malaysia is for **logging**.
Road building	Roads bring supplies and provide access to new mining areas, new settlements and energy projects. ● In Brazil, the Trans-Amazonian Highway stretches for some 4,000 kilometres through the rainforest. ● In Malaysia, logging companies use an extensive network of roads for heavy machinery and to transport wood.
Mineral extraction	**Mineral extraction** is a major cause of deforestation in many countries. ● In Brazil, minerals such as gold, bauxite and copper are mined extensively, causing huge scars in the landscape and polluting rivers. The largest iron ore mine in the world is at Carajás, which is worked by 3,000 people, 24 hours a day. ● Borneo has rich reserves of tin, copper and gold. Coal is an important source of energy, with 99% of Malaysia's supply in Borneo. Tin mining is common in Peninsular Malaysia.
Energy development	High rainfall creates ideal conditions for hydro-electric power (HEP), and there are several large dams and reservoirs. ● In Brazil, the Belo Monte dam (currently under construction) will block the Xingu River, a tributary of the Amazon, flooding more than 40,500 hectares of rainforest and displacing more than 15,000 people. ● In Malaysia, there are several dams supplying hydro-electric power (HEP) including the Bakun Dam, completed in 2011. At 205 metres, it is the highest dam in Asia (not counting China). It supplies energy for industries in Peninsular Malaysia. Drilling for oil and gas has started in Borneo.
Settlement and population growth	Settlements have developed to service the developments in the Brazilian and Malaysian rainforests, such as farming and mineral extraction. This has led to an increase in population.

Revision activity

For your chosen case study country, make a list or use a spider diagram to identify the causes of deforestation.

Exam tip

You only need to learn about **one** case study, i.e. Brazil or Malaysia.

What are the impacts of deforestation?

Impacts	Description
Economic development	• Mining and commercial farming generates employment and income for the government. • Taxes can be used to improve education and social conditions. • Oil palm, rubber and other commercial farming products provide raw materials for processing industries, increasing the value of products sold abroad. • Hydro-electricity provides cheap renewable power, boosting industrial development. • Improved **infrastructure** opens up new areas for economic development and settlement. • However, longer-term economic losses might be felt as forest ecosystems are destroyed and land and rivers become polluted. • The loss of biodiversity may reduce tourism.
Soil erosion	**Soil erosion** is a major problem in many countries. Once land is exposed by deforestation, the soil is much more vulnerable to the torrential tropical rain. Without the roots to bind it together, the loose soil can easily be eroded and washed away.
Climate change	• At the local scale, tropical rainforests retain and emit moisture to the atmosphere, maintaining high levels of humidity and contributing to the local water cycle. When large areas of rainforest are cut down, the climate becomes drier. • High levels of evaporation from leaves and branches cool the air. When the trees are cut down, this cooling process stops and the air becomes warmer. • Trees are important 'sinks' for carbon dioxide, a greenhouse gas responsible for increasing global temperatures. During the process of photosynthesis, carbon dioxide is absorbed and then stored by the trees. No trees, no storage of carbon dioxide! • When trees are burned, carbon is released back into the atmosphere, enhancing the greenhouse effect and contributing towards global warming.

Now test yourself

1 Compare the impacts on the rainforest of subsistence farming and commercial farming.
2 Why are tropical rainforests suitable for the development of hydro-electric power (HEP)?

TESTED

Figure 6.10 Ouro Verde Gold Mine

Exam tip

The impacts of deforestation can be applied to either Brazil or Malaysia. You can use some of the specific detail in the 'causes' table (page 31) to help focus your answer on your chosen country, for example by referring to a named hydro-electric power (HEP) scheme.

Exam practice

1 The Amazon rainforest stretches across parts of Brazil, Peru and Colombia. Use Figure 6.7 (see page 30) to calculate the proportion of the world's tropical rainforest biomass in the Amazon.　(2 marks)
2 Study Figure 6.9 (see page 30). Describe the pattern of changing rates of deforestation.　(4 marks)
3 With reference to a case study, discuss to what extent the economic benefits of deforestation outweigh the losses.　(6 marks)
4 Explain why deforestation can contribute towards climate change.　(6 marks)

 ONLINE

6.3 Management of tropical rainforests

What is the value of tropical rainforests to people and the environment?

Tropical rainforests are extremely important global ecosystems. They are of great value both to people and to the wider environment.

Value to people	Value to the environment
Resources – tropical rainforests are rich in reserves of wood, nuts and fruit as well as minerals. Everyday items such as bananas, cocoa and sugar come from tropical rainforests, along with spices such as vanilla and cinnamon.	Water – rainforests are important sources of freshwater. About 20% of the world's freshwater comes from the Amazon Basin.
Medicine – about 25% of all medicines come from rainforest plants, and more than 2,000 plants have anti-cancer properties. Less than 1% of rainforest plants and trees have been tested by scientists for their medicinal qualities.	Biodiversity – tropical rainforests contain 50% of the world's plants and animals, including thousands of different species.
Indigenous tribes – thousands of people live in harmony with the rainforest, with their lives depending on maintaining a healthy ecosystem. For example, the Achuar tribe in Peru number over 11,000, living in small communities relying on the rainforest for food, building materials and fuel.	Climate – known as the 'lungs of the world', rainforests contribute 28% of the world's oxygen. Moisture emitted through transpiration feeds into the water cycle and prevents the climate becoming too dry and hot. Evaporation of water from rainforests helps to cool the air (heat is extracted from the air during the process of evaporation).
Energy – high rainfall totals in rainforests create the potential for hydro-electric power (HEP). Electricity can provide much-needed light and power for local people. Local micro-hydro schemes can serve isolated communities.	Climate change – rainforests absorb carbon dioxide (an important greenhouse gas) from the atmosphere, acting as a 'carbon sink'. This helps to offset global warming.
Employment – rainforests can provide employment opportunities in tourism as guides or stewards. Other opportunities exist in construction, farming and mining.	Soil erosion – rainforests shelter and bind together the tropical soils, preventing harmful soil erosion, which can silt up rivers and reservoirs.

Exam tip

Be prepared for an exam question that asks you to focus on either 'people' or the 'environment'.

Revision activity

Convert the information in the table to a spider diagram, with the values to people on the left-hand side and values to the environment on the right.

Now test yourself

How are tropical rainforests valuable for medicines, climate and energy?

TESTED

How can tropical rainforests be managed sustainably?

REVISED

Sustainable management involves establishing an environmental balance, enabling the rich resources of the rainforest to be used without causing any long-term damage to the ecosystem. Harmful practices such as clear-felling, illegal logging and heavily polluting mining need to be replaced by more sustainable, less destructive approaches.

Selective logging and replanting

This system, introduced in Malaysia as the Selective Management System in 1977, involves the **selective logging** of individual trees rather than the indiscriminate clear-felling that is so damaging to rainforests.

- Trees are selected by professionals, felled and extracted in such a way as to minimise damage to other trees.
- Officials monitor the logging to ensure that it is done legally and correctly.
- New trees are planted to ensure that the system is sustainable.

Conservation and education

Several countries have designated areas of rainforest as national parks, affording them protection from development. Some large international organisations have supported conservation projects in exchange for carrying out scientific research or using raw materials. The Swiss perfume company Givaudin works with Conservation International to conserve rainforests in Venezuela in exchange for obtaining tonka beans, used in the production of perfume.

Several international charities, such as the WWF, Birdlife International and Fauna and Flora International, support conservation and education programmes, training conservation officers and scientists, and promoting rainforest conservation in schools.

Ecotourism

Ecotourism is often encouraged in tropical rainforests and it has become widespread in countries such as Costa Rica, Belize and Malaysia. It is a sustainable form of tourism that focuses on the natural environment and has a very low impact. It supports local communities, employing indigenous people as guides, and often providing education or social improvements. By focusing on nature, ecotourism offers local people and governments a financial return for preserving rainforests, rather than cutting them down.

International agreements about the use of hardwoods

There are a number of international agreements and strategies to control the use of hardwoods:
- The Forest Stewardship Council (FSC) promotes sustainable management by approving timber from sustainable sources. Suppliers are encouraged only to buy wood with the FSC stamp. This is a form of environmental quality control.
- The International Tropical Timber Agreement (2006) restricts the trade in hardwoods by only marking timber with a registration mark if it is from a sustainably managed forest.

Debt reduction

Most countries with tropical rainforest are less developed countries (low income countries, LICs, or **newly emerging economies, NEEs**). To promote development, some have taken out sizeable loans which they now find hard to repay. Some high income countries have agreed to write off debts (**debt reduction**) in return for rainforests being protected. For example, in 2010 the USA agreed to convert a Brazilian debt of £13.5 million into a fund to protect areas of rainforest. This is called 'debt for nature swapping'.

Now test yourself

1 What is the meaning of the term 'sustainable' in the context of rainforest management?
2 How is selective logging and replanting a good example of sustainable management?

TESTED

Exam practice

1 Describe the value of tropical rainforests to the environment. (6 marks)
2 Explain how ecotourism can be an effective strategy in the sustainable management of tropical rainforests. (4 marks)
3 'International co-operation is the only way to protect rainforests in the future.' Do you agree with this statement? (6 marks)

ONLINE

7 Hot deserts

7.1 Characteristics of hot deserts

What are the physical characteristics of hot deserts?

Most of the world's **hot deserts** are found in the subtropics between 20 degrees and 30 degrees north and south of the Equator. The Tropic of Cancer in the northern hemisphere and Tropic of Capricorn in the southern hemisphere pass through most hot deserts.

- Aridity – the primary characteristic of a hot desert is its dryness or aridity. Hot deserts have total annual rainfall below 250 millimetres per year. This is caused by the dominance of extensive belts of high atmospheric pressure existing at these latitudes (see page 8). Sinking air prevents the formation of cloud and rain, resulting in the very dry conditions.
- Heat – as the name implies, summer temperatures of hot deserts rise well above 40°C. Did you know that there are also 'cold deserts' in the Arctic and Antarctic where precipitation is very low?
- Landscapes – deserts are often thought of as vast areas of towering sand dunes, but most deserts are rocky, desolate places with isolated thorny bushes and cacti.

Now test yourself

Can you name a hot desert in each of the five continents, excluding Europe and Antarctica?

TESTED

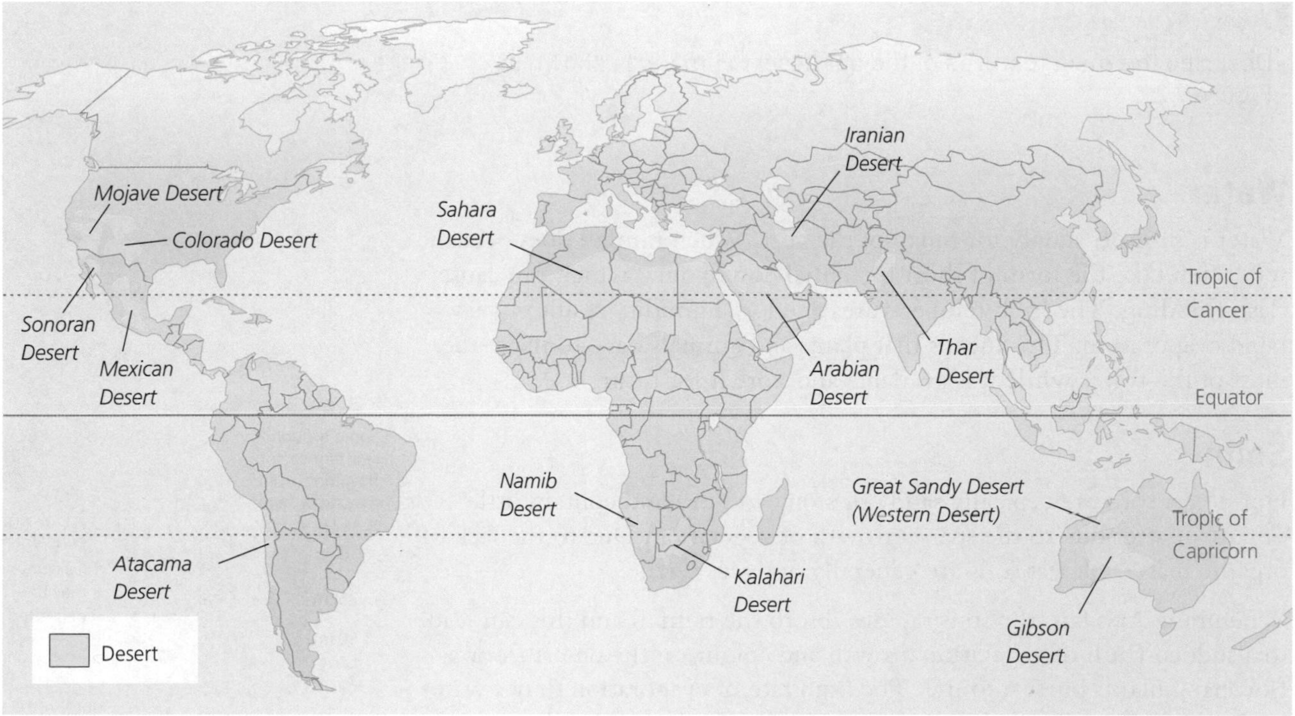

Figure 7.1 Global distribution of the world's hot deserts

Climate

Figure 7.2 shows the main characteristics of the hot desert climate. The main driving force determining the climate is the presence of the sub-tropical high-pressure zone either side of the Equator. This is part of the global atmospheric circulation system (see page 8). Here air sinks towards the ground, suppressing any rising air that would subsequently cool and condense to form rain.

The predominantly clear skies account for the very high daytime temperatures but quite cool night-time temperatures, as heat is allowed to escape to the atmosphere. In the winter, hot deserts can experience the occasional frost and even light snowfall!

The climate is the most important physical characteristic affecting all aspects of the hot desert ecosystem and accounting for the strong interdependence that exists in these hostile environments.

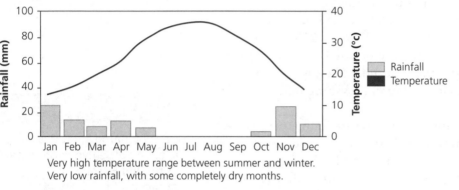

Very high temperature range between summer and winter.
Very low rainfall, with some completely dry months.

Figure 7.2 A hot desert climate in Salah, Algeria

Now test yourself

TESTED ☐

Describe the main features of the hot desert climate as shown in Figure 7.2.

Water

Water is in short supply for much of the year. When rainfall does occur it will often take the form of short torrential downpours, which can cause flash flooding. The high temperatures and low humidity result in very rapid evaporation. This means that plants and animals have to make the most of the water while it is available and store it for future use.

Soils

Hot desert soils are typically sandy or stony and dry and contain little organic matter due to the sparse growth of vegetation. Due to the lack of organic matter, desert soils are generally not very fertile.

When rain does fall, the soils rapidly absorb the rainfall and this can lead to a sudden flush of vegetation growth and colour on the desert floor as flowering plants burst into life. The high rate of evaporation draws water to the surface, leaving behind salt deposits that form an infertile crusty surface. This process is called salinisation.

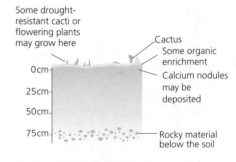

Figure 7.3 A desert soil profile

Plants and animals

Despite the low level of **biodiversity** in hot deserts, there is a surprising number of species of plants and animals that have adapted to the harsh environmental conditions.

Plants have adapted in the following ways:

- Leaves are thick and waxy to retain water and reduce water loss through transpiration. Many plants and shrubs have developed needles rather than leaves, such as cacti (Figure 7.4).
- Plants have developed ways of storing water in their roots, leaves or stems (e.g. cacti) – these plants are called succulents.
- Some plants and trees have extensive root networks, extending to 10 metres or more to tap into groundwater deep below the surface.
- Many plants have rapid life cycles triggered by occasional rainfall events, germinating, flowering and setting seed in a matter of days.

Animals have adapted in the following ways:

- Many rodents are nocturnal, hunting during the cool nights. These animals often have large eyes to help them see in low light conditions.
- Animals often live in burrows, enabling them to stay cool during the daytime.
- Some animals, such as the jerboa, have large ears to enable heat loss to occur.
- Kangaroo rats obtain water from their food, live in burrows during the day and have highly efficient kidneys enabling them to produce little urine. They also do not perspire.
- Camels – described as the 'ships of the desert' – are supremely well adapted to living in hot deserts (Figure 7.5).

Needles instead of leaves, reducing the surface area in contact with the air, minimising loss by transpiration.

Large capacity to store water in fleshy stems.

Large network of roots to absorb water rapidly after rainfall.

Figure 7.4 Cactus adaptations for survival in hot deserts

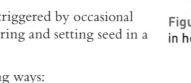

> **Now test yourself**
> TESTED ☐
>
> Describe how cacti, kangaroo rats and camels are well adapted to living in hot desert environments.

People

While hot desert interiors tend to be bleak and desolate, desert fringes (semi-arid regions) are sparsely vegetated and used by people mainly for livestock farming. Tribes such as the Bedouin in Jordan have adapted their lifestyles to cope with the harsh conditions:

- They live in large open tents, which keep them cool during the day but warm at night.
- Traditionally, food is cooked slowly within the warm sandy soils.
- Head scarves (keffiyeh) traditionally worn by men provide protection from the Sun, warmth at night, and can cover the mouth and nose during sand storms.

Irrigation can transform deserts into agriculturally productive areas, for example along the banks of the River Nile in Egypt.

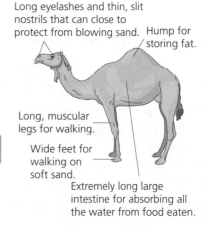

Long eyelashes and thin, slit nostrils that can close to protect from blowing sand.

Hump for storing fat.

Long, muscular legs for walking.

Wide feet for walking on soft sand.

Extremely long large intestine for absorbing all the water from food eaten.

Figure 7.5 Camel adaptations to living in hot deserts

> **Revision activity**
>
> Draw a simple sketch of Figure 7.4, adding located labels to describe the adaptations to hot desert environments.

What interdependence exists in hot desert environments?

REVISED

Interdependence between the various components of the hot desert ecosystem is shown by:

● the existence of complex food webs (Figure 7.6) whereby energy and nutrients gained from water, soils and vegetation are transferred between different species
● the sustainable coexistence of people, plants and animals in fragile semi-arid environments
● the adaptations of plants and animals to the climate and soil characteristics
● the potential damage to the ecosystem inflicted by people engaged in unsustainable practices, such as **overgrazing**.

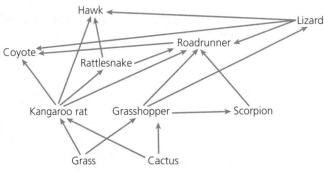

Figure 7.6 Simplified food web for the Western Desert, USA

How biodiverse are hot deserts?

REVISED

Hot deserts are considered to have low levels of biodiversity. This reflects the low rainfall and lack of available water. However, rich biodiversity does exist in pockets where water is present, for example close to an oasis where underground springs emerge at or close to the ground surface. Rich biodiversity can also exist close to seasonal lakes or streams.

Hot desert ecosystems are extremely fragile. The slightest change in biotic or abiotic conditions can have disastrous consequences – human activities are most likely to threaten biodiversity in these localities.

The gueltas of Mauritania – biodiversity under threat

Gueltas are small mountain rock pools found in Mauritania, and are typically about 100 square metres in size. They form isolated hotspots of biodiversity, attracting nearly 80 per cent of Mauritania's endemic species (including mammals, amphibians, bats and invertebrates), particularly after the seasonal rains that fall between July and September.

However, severe droughts in the 1970s resulted in some northern pools drying up completely. This led to the once nomadic herders in the region becoming more sedentary farmers close to the remaining pools. The increased pressure on this limited water resource, together with contamination from animals, has led to a reduction in biodiversity and threatens the entire fragile ecosystem.

Now test yourself

Describe how the food web in Figure 7.6 illustrates interdependence between components of the desert ecosystem.

TESTED

Revision activity

The example of gueltas in Mauritania illustrates both biodiversity issues and interdependence. Draw a simple flow diagram to show how biodiversity at these water sources is threatened.

Exam practice

1 Study Figure 7.1. Describe the distribution of the world's hot deserts.
(4 marks)
2 What are the physical characteristics of hot deserts? (4 marks)
3 Describe how animals have adapted to living in hot deserts. (6 marks)
4 Assess the importance of interdependence between abiotic and biotic components of the hot desert ecosystem.
(9 marks)

ONLINE

Exam tip

The term 'interdependence' is used in the specification so be prepared to adapt your knowledge of cold environments to focus on this term. Consider how the different physical characteristics interrelate with one another.

7.2 Case study: development of hot deserts

What are the development opportunities in hot deserts?

We are going to consider developments in two hot deserts, but you only need to learn about one of them!

- The USA's Western Desert stretches across several states in the south-west, including California, Arizona and New Mexico. It comprises three separate deserts: the Sonoran Desert, the Mojave Desert and the Chihuahuan Desert.
- The Thar Desert stretches across northwest India and into Pakistan. It covers around 200,000 square kilometres, about the same size as the USA's Western Desert. It is the most densely populated desert in the world.

> **Exam tip**
>
> Remember that you only need to learn one of the case studies, but you should do so thoroughly!

Development opportunity	Western Desert, USA	Thar Desert, India/Pakistan
Mineral extraction	Rich reserves of copper, uranium and coal. Copper is mined in the Sonoran Desert near Ajo, Arizona. Elsewhere, developments have been limited by environmental concerns, including the possibility of mining uranium (raw material for nuclear power) in the Grand Canyon.	The following are used within India and exported worldwide: ● Phosphorite – used to manufacture fertiliser ● Gypsum – used to make cement and plaster for construction ● Kaolin – a white clay used in making paper.
Energy	Energy resources include: ● solar energy being developed extensively, e.g. Sonoran Solar Project in Arizona, which will supply electricity to 100,000 homes ● oil in Arizona, owned and operated by the indigenous Navajo people ● HEP generated from Lake Mead.	Renewable and non-renewable resources include: ● oil – a large oilfield has been discovered in Barmer ● coal – there are widespread lignite (low quality) coal deposits and a thermal power station at Giral ● wind – India's third-largest wind farm is the Jaisalmer Wind Park ● solar – huge potential due to the long hours of sunshine. Solar energy is used in water treatment at Bhaleri.
Farming	Irrigation enables commercial farming to thrive in sunny and hot conditions, e.g. Coachella Valley (California) produces vegetables, peppers and grapes (for wine) using water abstracted from underground aquifers. Elsewhere, irrigation canals are used for large-scale industrialised farming.	Mostly subsistence farming involving grazing animals, growing crops and foraging from fruit trees. Extensive irrigation made possible by the Indira Gandhi Canal enables commercial farming to thrive, growing wheat, maize and cotton.
Tourism	Many people visit the National Parks and **wilderness areas** (e.g. Joshua Tree and Grand Canyon) to enjoy the landscape, peace and solitude. Lake Mead and Lake Powell are popular for watersports and Las Vegas is an entertainment centre attracting 37 million visitors a year.	Tourism has increased recently, particularly from Pakistan. Several companies offer desert safaris and visits to Jaisalmer. Ecotourism is popular with small groups taking camel treks into wilderness areas or to visit oases.

> **Revision activity**
>
> For your chosen case study, construct a spider diagram to summarise the development opportunities.

What are the challenges of developing hot desert environments?

Hot deserts are hostile and often very remote environments, presenting challenges for development.

Water supply

For the Western Desert, water transfer, particularly from the Colorado River (Figure 7.7), has supplied drinking water and irrigation since the Hoover Dam was constructed in 1935. Piped water now supplies homes, farms and even golf courses. Water supply could be a major problem in the future as demand soars in cities such as Phoenix, Arizona. Temperatures are forecast to rise as a result of climate change.

In the Thar Desert, with low rainfall and high rates of evaporation, a secure water supply is crucial for economic development.

- Traditionally water is stored in natural ponds called tobas, used by farmers in remote areas.
- Most rivers are erratic but settlements tend to cluster along their edges. Climate change could make river flow less reliable.
- The Indira Gandhi Canal (constructed in 1958) has transformed the desert, providing drinking water and irrigation.

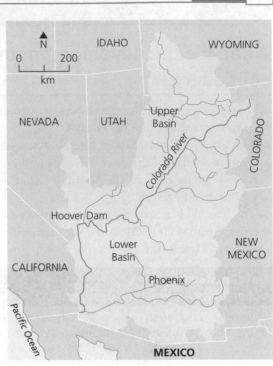

Figure 7.7 The Colorado River's water supply

Inaccessibility

In the Western Desert, the relatively low density population has resulted in a lack of surfaced roads through the desert. Access is limited away from the main cities. Some major highways have been constructed linking the major cities, such as Route 70 through Utah, and Route 66, which connects Chicago with California through the Western Desert.

In the Thar Desert, despite the relatively high population density, there is a limited road network due to the vast distances and high maintenance costs. Sand can easily blow over the roads and tarmac can melt in the extreme heat. Camels are the traditional form of transport.

Extreme temperatures

For both deserts, while the average temperature is 27°C, summer temperatures can soar above 50°C. This creates many problems:
- Work outside is very hard, especially for farmers who have to work during the day.
- High temperatures lead to high rates of evaporation and water shortages.

Plants and animals have to adapt to cope with the high temperatures or be provided with shelter (particularly in the Thar Desert). Traditional houses in the Western Desert have thick earth walls to keep homes cool during the day but warm at night. Whitewashed walls reflect sunlight.

Now test yourself

Why are extreme temperatures a challenge for development?

Exam practice

1 Use a case study to describe the opportunities for developing mineral extraction and energy in hot deserts.   (6 marks)
2 'Water supply is the most significant challenge facing development in hot deserts.' With reference to a case study, to what extent do you agree with this statement?   (6 marks)

Exam tip

Make sure you focus on all parts of the exam question. So for Question 1, write about mineral extraction **and** energy in a balanced way. For Question 2, make sure you express a point of view and justify it.

7.3 Desertification

What is desertification?

Desertification is quite simply where land turns into desert. This is a huge global environmental issue affecting vast swathes of land, particularly semi-arid areas (Figure 7.8). These fragile areas are easily damaged and degraded both by natural factors, such as climate change, and human actions involving poor land management. An estimated 1 billion people live in areas at risk from desertification.

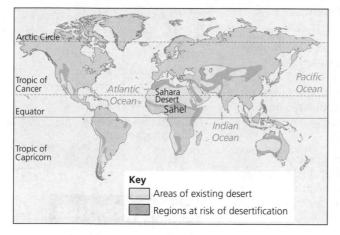

Figure 7.8 Areas at risk of desertification

What are the causes of desertification?

Climate change

Semi-arid areas are one of the most fragile ecosystems in the world. They depend on relatively low rainfall totals for their survival.

Since the 1970s, the Sahel region of West Africa has experienced significant reductions in annual rainfall, with just the occasional year receiving higher than average rainfall. Some scientists believe that climate change may be disrupting the normal patterns of rainfall, and causing more droughts.

Population growth

In the Sahel region, the population has increased from 30 million people in 1950 to almost 500 million today. This is expected to double by 2050. As demand for food has increased, the land is put under enormous pressure, leading to vegetation destruction, soil erosion and ultimately desertification. The problem has been made worse by people migrating away from drought-stricken regions or areas affected by war or natural disaster.

Removal of fuelwood

Millions of people living in semi-arid regions depend upon wood as their primary source of fuel. When land becomes stripped of its trees, it is vulnerable to erosion by wind and rain and it can quickly become degraded and turn to desert.

Overgrazing

Overgrazing by animals such as goats, sheep and cattle results in vegetation being stripped from an area and soil being left bare and vulnerable to erosion. Overgrazing often happens because the available pasture land is reduced, by natural causes (such as drought) or human factors (such as political conflicts, or restrictions on the free movement of nomadic herders). Population pressure can also lead to overgrazing.

Over-cultivation

Over-cultivation is similar to overgrazing, but takes place when land is over-cultivated for growing crops, exhausting the soil of its nutrients. Soils in semi-desert regions are not very fertile and contain little organic matter. When soils are over-cultivated, often in response to population pressure or restrictions in available land, the soil becomes dry, dusty and infertile. It is then prone to soil erosion.

Soil erosion

When vegetation is stripped or killed, the soil is left bare, to be baked hard by the Sun. When rainfall occurs, it washes over the hard surface, eroding rills and gullies and washing away the topsoil. This leaves behind a very infertile sub-soil that cannot be used for anything. Soil can take thousands of years to form but can be eroded away in just a matter of hours.

> **Revision activity**
>
> Find a photo of desertification that illustrates some of the causes described above. Add text boxes to summarise each of the causes.

> **Now test yourself**
>
> 1 What is desertification, and where in the world is it a problem?
> 2 Explain how overgrazing and over-cultivation can lead to desertification.
>
> TESTED

What strategies can reduce the risk of desertification?

REVISED

On a global scale, there is no simple solution to addressing climate change. If there is a link between climate change and desertification it is highly complex and little understood. For this reason, strategies for reducing desertification tend to focus on addressing the more obvious human causes.

Strategies	How these reduce the risk of desertification
Water and soil management	Water and soil management addresses the problem of intense rainfall events washing away loose soil and causing soil erosion. It often involves water storage and attempts to control the surface flow of water. One common strategy used in parts of the Sahel (e.g. Burkina Faso) and the Middle East (e.g. Jordan) involves constructing a series of low rock walls called bunds. These are deliberately constructed to follow the contours of the land, interrupting the downslope flow of surface water. Any soil carried by the water is deposited on the upslope side of the walls, creating a reasonable thickness of soil that can then be cultivated.
Tree planting	Trees are very effective in preventing soil erosion and desertification. They act as an umbrella, protecting the soil from the direct impact of torrential rain and providing shade for seedlings. Their roots help to bind the soil together, preventing it being washed or blown away. ● In the Thar Desert, India, the *Prosopis cineraria* tree has been planted to address the problem of desertification. It is well adapted to desert conditions and, if managed correctly, will provide foliage and seed pods for animals to eat and wood for firewood and building. ● In 2007, the African Union launched a project called the 'Great Green Wall', an ambitious plan to plant trees across the southern edge of the Sahara Desert to reduce desertification and address issues such as water management and food security. Some 21 African countries are involved, and already 15 million hectares of land have been restored in Ethiopia alone.
Use of appropriate technology/ intermediate technology	Appropriate technology has been used in a variety of ways to address desertification: ● The use of low walls (bunds) to manage water is a good example as it makes use of local materials, basic tools and transport and is based on a simple idea. Community-led, it makes use of local people who work together for a common aim. ● Alternative cooking devices that use efficient stoves (often supplied by charities such as Practical Action) burn small amounts of wood or charcoal, reducing the quantity of fuelwood required.

Now test yourself

1 What is appropriate technology?
2 How can appropriate technology be used in water and soil management?

TESTED

Exam practice

1 What is overgrazing? (2 marks)
2 Explain how population growth can cause desertification. (4 marks)
3 Evaluate the strategic options for reducing the risk of desertification. (9 marks)

ONLINE

Exam tip

Each of the headings used to describe the causes of desertification (page 41) and strategies to reduce risk (table on the left) are in the specification, so be prepared to answer a question on them.

8 Cold environments

8.1 Characteristics of cold environments

What are the physical characteristics of cold environments?

Cold environments experience very low temperatures for long periods of time. The most extreme environments, such as Antarctica, experience temperatures well below freezing throughout the year, creating almost impossible conditions to support life.

Figure 8.1 shows the distribution of the world's cold environments. You need to focus on two of them: polar and tundra.

- Polar – this is the most extreme cold environment and includes Antarctica and much of Greenland. The extreme cold and permanent darkness during the winter months combine to make this one of the most inhospitable environments on Earth.
- Tundra – this cold environment borders the polar region to include parts of Canada, Alaska (USA), Northern Europe and Russia and the far southern tip of South America. Despite the cold winters, conditions are less harsh, enabling life to thrive, especially in the summer.

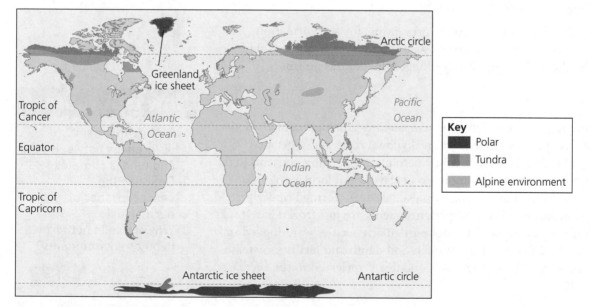

Figure 8.1 Distribution of the world's cold environments

Now test yourself

Describe the distribution of cold environments.

Climate

Polar climate	Tundra climate
• Temperatures remain below freezing (0°C) throughout the year. • Temperatures in the winter can dip below −50°C. • Precipitation (snowfall) is low; this is a 'cold desert'. • Winters are permanently dark, with no sunshine for several months of the year.	• There is a significant annual temperature range, from well below freezing (−30°C or lower) in the winter to well above freezing in the summer. • Precipitation is higher, especially in the summer when the warmer air can hold more moisture. Coastal regions can receive heavy snowfall in the winter. • Winter months experience near or total darkness.

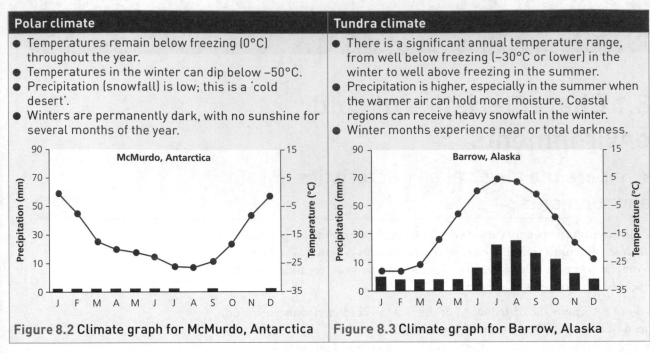

Figure 8.2 Climate graph for McMurdo, Antarctica

Figure 8.3 Climate graph for Barrow, Alaska

Permafrost

One of the unique characteristics of polar and tundra environments is the existence of **permafrost**, permanently frozen ground. In tundra environments, the upper surface layer – usually no more than a few centimetres – melts in the summer. Known as the active layer, this can become mobile on slopes, slowly slumping downhill through a process called solifluction (literally meaning 'soil flow').

Permafrost is hard and impermeable, preventing any soil formation processes from operating and acting as a barrier to water transfer. This explains why the active layer is waterlogged during the summer.

Soils

Soil formation requires relatively high temperatures and high rainfall. Under these conditions, weathering actively breaks down rocks and vegetation grows, dies and is then broken down by decomposers, adding nutrients to the soil. In cold environments, conditions do not favour soil formation!

- Polar soils – almost non-existent unless they were formed under past climatic conditions. Most polar environments are just frozen bare rock.
- Tundra soils – tend to be thin, not particularly fertile, waterlogged in the summer and frozen in the winter. Soil depth and fertility increase with distance away from the Poles, and this is reflected in the increased **biodiversity**.

Now test yourself

1 Outline the main differences between polar and tundra climates.
2 Describe the characteristics of permafrost.
3 Why are soils not formed in polar environments?

TESTED ☐

Plants

Apart from a few isolated patches of moss and lichen, there are no plants in polar environments. Plants are much more prolific in tundra environments where they exhibit a number of adaptations:

- Shallow root systems enable them to access nutrients and water close to the surface within the active layer. Roots cannot penetrate the permafrost.
- Low-growing 'cushions' of plants retain moisture and provide shelter from the strong drying winds, e.g. the Arctic willow.
- The bearberry plant has thick, stunted stems to help it withstand strong winds, and hairs on its stems to retain warmth. Its small, leathery leaves retain moisture.
- Mosses can cope with waterlogged conditions in the summer and will also survive periods of winter drought.
- Flowering plants such as the snow buttercup and Arctic poppy have rapid life cycles, enabling them to flower and set seed very quickly during the short summers.

Animals

In polar regions, animals have to survive on the available food supply, which is generally in the oceans. Polar bears survive mostly by eating seals, whereas, in the Antarctic, penguins feed on fish, krill and crustaceans. Food chains are short and food webs very basic.

In tundra regions, with the presence of plants, food webs are more complex and biodiversity is greater (Figure 8.4).

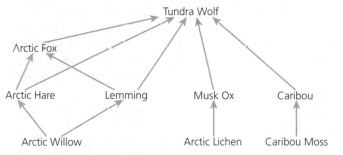

Figure 8.4 Tundra food web

Animals living in these harsh environments display a number of adaptations:

- Arctic foxes and Arctic hares – thick fur on their bodies and feet keep them warm; their fur becomes white (good camouflage) in the snowy winter. Arctic foxes can survive on a wide variety of foods, which helps with the cold winters, when food is scarce.
- Musk oxen – two layers of fur help to insulate them; wide hooves enable them to walk on snow or waterlogged ground.

People

There are no permanent settlements in polar environments, apart from small groups of research scientists in Antarctica. However, indigenous people do live in tundra environments, particularly at the coast where the sea provides a rich source of food.

The Inuit are a group of indigenous peoples that inhabit parts of Greenland, Alaska and Canada. Unable to grow crops or rear animals, they survive as hunter-gatherers by fishing and hunting seals and even whales. In the summer they gather berries and tubers from tundra plants. They dress warmly (traditionally using animal skins) to survive the cold and live in well-insulated houses. Travel is usually by sleds pulled by husky dogs or increasingly by snowmobile.

Now test yourself

1 For a named plant, describe how it has adapted to cope with living in a cold environment.
2 How are Arctic foxes and musk oxen well adapted to living in polar environments?

TESTED

Revision activity

Find a photo of a polar or tundra animal and add labels to describe how it is well adapted to the environment.

Exam tip

When describing plant or animal adaptations, try to refer to named species. This will help you to access a higher level mark in the exam.

Now test yourself

How are the Inuit people well adapted to living in cold environments?

TESTED

How is a cold environment interdependent?

REVISED

Interdependence is essential for survival in harsh cold environments. It can be demonstrated in a number of ways:

- Indigenous peoples depend on animals such as seals and whales for food, clothing and heating/lighting oil. The snow and ice are used in constructing traditional shelters (igloos).
- Plants form dense cushions on the ground to retain moisture and heat and to provide shelter from the strong winds.
- The bright red berries of the bearberry attract owls and other birds, which eat the berries and therefore disperse the seeds.
- Arctic birds use moss to provide warmth and shelter in their nests.
- In tundra environments, complex food webs are good illustrations of interdependence (see Figure 8.4).

How biodiverse are cold environments?

REVISED

Tundra environments exhibit considerable biodiversity (Figure 8.4), with thousands of species of lichens, flowering plants and insects. Most species are heavily dependent on the short mild summers when the surface permafrost (the active layer) thaws and plants burst into life, providing food for other species further up the food chain. A number of issues may affect biodiversity in the future:

- Climate change could create longer, warmer summers in the Arctic, melting permafrost and potentially increasing biodiversity as conditions become less extreme. However, numbers of certain species may increase rapidly, causing an imbalance within the ecosystem.
- Melting permafrost releases carbon to the atmosphere, further enhancing the greenhouse effect and accentuating the rate of climate change.
- Increased use of Arctic regions by people, for example in extracting oil and other resources, may cause pollution and upset the fragile ecosystem. Oil extraction in Siberia has already had catastrophic impacts on local tundra ecosystems.
- Warming of the oceans might affect the distribution of nutrients (cold waters tend to be rich in nutrients), which in turn may affect concentrations of krill, fish and crustaceans.

Group	Number of species
Mammals	75
Birds	240
Insects	3,300
Flowering plants and shrubs	1,700
Mosses	600
Lichens	2,000

Now test yourself

TESTED

How may climate change affect biodiversity in cold environments?

Exam practice

1 What is permafrost?
(2 marks)
2 How have plants adapted to cope with living in cold environments? (4 marks)
3 'Survival in cold environments depends on interdependence.' To what extent do you agree with this statement?
(9 marks)

ONLINE

Exam tip

The term 'interdependence' is used in the specification, so be prepared to adapt your knowledge of cold environments to focus on this term. Try to use examples of interdependence, such as those listed above, involving the physical environment, plants, animals and people.

8.2 Case study: development of cold environments

What are the development opportunities in cold environments?

REVISED

We are going to consider developments in two cold environments, but you only need to learn about one of them!

- Alaska – the USA's most northerly state covers a vast area, some 2 million square kilometres. While the southern half experiences relatively mild conditions, northern Alaska is a tundra environment underlain by thick permafrost.
- Svalbard – these Norwegian islands, situated in the Arctic Ocean, are the most northerly permanently inhabited islands in the world. Some 60 per cent of the land area is a true polar environment, comprising bare rock and glaciers. The rest is tundra, underlain by thick permafrost.

> **Exam tip**
>
> Remember that you only need to learn one of the case studies, but you should do so thoroughly!

Development opportunity	Alaska, USA	Svalbard, Norway
Mineral extraction	Alaska has a rich resource base. In the 1800s there was a 'gold rush' to the state. Gold still accounts for 20% of Alaska's mineral wealth. Other minerals mined include silver, zinc and lead.	Vast reserves of coal, much of which is exported to Russia. Mining employs over 300 people. A new coal mine opened near Svea in 2014.
Energy	Alaska has vast reserves of oil and gas in the far north of the state. The industry employs 100,000 people and accounts for a third of the state's annual income. Oil is transported some 1,300 kilometres from Prudhoe Bay in the north to the port of Valdez in the south by the trans-Alaskan pipeline. From Valdez, the oil can be transported by tanker through ice-free waters. There are significant conservation concerns about developments in the far north's pristine wilderness environment.	Coal is used to power Longyearbyen power station, which supplies all of Svalbard. There are thought to be extensive reserves of oil and gas in Svalbard's coastal waters. In 2011, Norway announced a twenty-year development plan. However, there are major conservation concerns about exploiting oil and gas. For the future, there is potential for the development of geothermal energy as Svalbard is close to the North Atlantic Ridge (constructive plate boundary).
Fishing	Alaska's rivers and coastal waters are rich in fish, including salmon, trout and several species of whitefish. Nearly 80,000 people are employed in the industry, which is worth $6 billion to the economy annually. Indigenous people use fish to provide food, oil and items of clothing (bones).	The nutrient-rich cold waters around Svalbard are one of the richest fishing waters in the world. They are important breeding and nursery grounds for fish, with over 150 species. Conservation and sustainable management are very important.
Tourism	Alaska's spectacular wilderness environment – mountains, glaciers, national parks – attracts up to 2 million tourists a year. It is one of Alaska's main employers. Cruises are particularly popular, accounting for 60% of the summer visitors. Adventure tourism is a major growth area.	Tourism has increased considerably in recent years (hiking, kayaking, snowmobile safaris, Northern Lights). Longyearbyen is the main centre for tourism, with some 70,000 visitors a year (including 30,000 cruise passengers). Longyearbyen's harbour has been enlarged to cope with the greater number of cruise ships. Tourism employs about 300 people.

> **Now test yourself**   TESTED
>
> For your chosen cold environment case study, describe the opportunities for tourism.

> **Revision activity**
>
> For your chosen case study, construct a spider diagram to summarise the development opportunities.

What are the challenges of developing cold environments?

Cold environments are hostile and often very remote environments, presenting challenges for development.

Challenge	Alaska, USA	Svalbard, Norway
Extreme temperature	Winter temperatures in the north can fall well below −30°C. This creates very hostile working conditions for people involved in the oil industry. In the winter the sea freezes and road conditions become treacherous.	Winter temperatures in Longyearbyen can fall below −30°C. In the polar regions it is even colder! This makes outside work extremely challenging and potentially dangerous. Those people working in the mines have to cope with very demanding conditions.
Inaccessibility	Alaska is a very remote region, accessible mostly by plane or ship. Road transport through Canada is lengthy and difficult in the winter. Anchorage is the main international gateway by plane, with local services operating to smaller towns. Individuals rely upon 4x4 vehicles or snowmobiles in the winter.	Svalbard is very remote and can only be reached by plane or ship. There is one international airport, at Longyearbyen, with flights from Norway and Russia. There is a very limited road network (about 50 kilometres), mostly around Longyearbyen. Transport mostly involves boat or snowmobile.
Buildings and infrastructure	The extreme cold, high winter snowfall and presence of permafrost combine to create challenges for building and infrastructure. To prevent melting of permafrost and subsidence, roads are constructed on raised gravel beds to prevent heat transfer. Domestic services (e.g. water, sanitation) are provided in above-ground insulated 'utilidors'. Airport runways are painted white to reflect sunlight and prevent them heating up.	People involved in construction (roads, buildings, harbour extension) have to cope with very challenging weather conditions (extreme cold and winter darkness). Buildings are very well insulated. The frozen ground (permafrost) provides firm foundations but care must be taken to prevent thawing and subsidence. Gravel roads, raised above the ground surface (to prevent heat transfer), are relatively cheap to maintain. Domestic services (e.g. water, sanitation) are raised off the ground in insulated pipes so they can be serviced and to prevent possible melting of permafrost.

Now test yourself

Why are extreme temperatures a challenge for development?

Exam practice

1 Use a case study to describe the opportunities for developing energy and fishing in cold environments. (6 marks)
2 'Inaccessibility is the most significant challenge facing development in cold environments.' With reference to a case study, to what extent do you agree with this statement? (6 marks)

Exam tip

Make sure you focus on all parts of the exam question, i.e. in Question 1 write about energy and fishing in a balanced way. In Question 2, make sure you express a point of view and justify it.

8.3 Protection of cold environments as wilderness areas

What is a wilderness area?

Wilderness areas are unspoilt, remote parts of the world, such as hot deserts, mountains and cold environments. By definition, they have largely escaped human development and remain natural and undisturbed.

Many of the world's cold environments can be considered wilderness areas due to their remoteness and extremely hostile physical conditions. Such areas include Antarctica, as well as parts of Alaska, Greenland, Siberia, Iceland and Svalbard.

Why should wilderness areas be protected?

There are several reasons why cold environment wilderness areas should be protected:

- Polar and tundra regions are fragile environments. They develop extremely slowly and take many years to recover from damage inflicted by people, such as pollution (Figure 8.5), mining or transportation (especially off-road driving). It is said that even a footprint on tundra vegetation will remain for ten years!
- Scientists need to conduct research in unspoiled environments to understand global processes of change. For example, research in the Antarctic has been invaluable in studying climate change.
- Some wilderness areas are inhabited by indigenous people, for example the Inuit, whose culture and very survival depend on the protection of the natural world.
- Cold environments provide important habitats for many species of plants, animals and birds, such as penguins, polar bears and the Arctic fox.
- There is a global moral responsibility to retain some wilderness areas that reflect the natural world without human interference.

Figure 8.5 Polluted river, Siberia, Russia

1 What is the evidence in Figure 8.5 that the cold environment is being damaged by human activity?
2 Why should this environment be protected?

What strategies can be used to maintain cold environments?

The following are all strategies to manage cold environments by balancing economic development with conservation.

Technology

Technology can provide environmentally friendly solutions to some of the challenges of developing cold environments. These can include the use of insulated pipes contained within utilidors to carry domestic services to people's houses.

On a larger scale, the trans-Alaskan pipeline is a superb example of how technology helps development without damaging the environment. Constructed in 1974, the 1,300-kilometre pipeline transports oil from Prudhoe Bay (where winter sea ice prevents the use of tankers) to the Pacific Ocean port of Valdez.

- The pipeline is insulated to retain the heat of the oil and prevent melting of the permafrost.
- It is raised well above the ground so herds of wild animals such as caribou can migrate across the area.
- Pumping stations keep the oil flowing over mountainous areas and across river valleys.
- Special slides enable the pipeline to accommodate movement during an earthquake and prevent fracturing.

International agreements

The continent of Antarctica is often described as the world's 'last great wilderness'. Despite its rich reserves of valuable minerals, it has remained undeveloped due to an international agreement signed in 1959 called the Antarctic Treaty, which came into force in 1961. This treaty

- stipulates that Antarctica should only be used for peaceful purposes, with all military activities banned
- promotes international co-operation in scientific research; there are several research stations in Antarctica and much of the evidence for climate change has come from here
- bans the disposal of nuclear waste
- encourages tourism but applies strict controls in terms of numbers and landing sites, to minimise impact.

Strategies to maintain cold environments

Action by governments

The US government has been involved in the protection of Alaska since oil was discovered here in the 1960s:

- Fisheries and marine habitats are monitored and protected by the National Oceanographic and Atmospheric Administration (NOAA).
- The Department of the Interior has created the Western Arctic Reserve in the north of Alaska to protect the area from oil and gas developments, and conserve the habitats of animals such as caribou and migrating birds.

Conservation groups

The worldwide Fund for Nature (WWF) works with governments, businesses and local communities across the Arctic to protect the region's biodiversity. In 1992 it launched the WWF Arctic Programme to work with governments on issues such as climate change, shipping, oil and gas and polar bears. Experts in several countries promote their vision:

'Effective international stewardship will shield the Arctic from the worst effects of rapid change by promoting healthy living systems to the benefit of local peoples and all humanity.'

Specific projects include:

- supporting scientific research to protect endangered species such as polar bears and Greenland sharks
- working with indigenous communities, oil companies and governments to promote sustainable development
- monitoring and seeking protection of threatened ecosystems.

Figure 8.6 Strategies to manage cold environments

Now test yourself

How has technology enabled oil to be transported in Alaska?

Exam practice

1 What is meant by a 'fragile environment'? (2 marks)
2 Explain why fragile environments need protecting. (6 marks)
3 Evaluate the strategies used to balance the needs of economic development and conservation in cold environments. (9 marks)

ONLINE

Revision activity

Construct a simple table to outline the four key strategies for managing cold environments.

Exam tip

In answering Question 3, you must focus on the command word 'evaluate', which requires you to write about the pros and cons of the different strategies.

9 UK physical landscapes

9.1 UK landscapes

How diverse is the landscape of the UK?

REVISED

The **landscape** of the UK varies greatly. There are spectacular, jagged mountain ranges in Scotland, rolling hills and valleys in much of central and southern England and extensive flat plains in East Anglia.

The term 'relief' can be used to describe the physical landscape, whereas the term 'landform' is used to describe an individual feature of a landscape, such as a waterfall.

The physical landscape – or relief – of the UK reflects the geology, the nature of the underlying rocks. There is a huge range of rock types exposed in the UK, which explains our varied scenery of uplands and lowlands.

Where are the UK's uplands?

REVISED

- Tough igneous and metamorphic rocks, such as granite and slate, form most of the mountains, such as the Grampians and the northwest Highlands in Scotland and the Lake District in northwest England.
- Resistant limestone (a sedimentary rock) forms the Pennine Hills, described as the 'backbone' of England.
- Ancient granite (igneous rock) forms the moors of southwest England such as Dartmoor and Bodmin (not Exmoor, which is made of sandstone).
- Bands of chalk and limestone (sedimentary rocks) form extensive asymmetrical ridges called escarpments in southern England. These include the Chilterns, the Cotswolds and the North and South Downs.

> **Exam tip**
>
> When writing about relief, refer to heights above sea level, the steepness of slopes (gradient) and landforms such as hills and valleys. Be prepared to use an OS map extract to do this.

Where are the UK's lowlands?

REVISED

- Much of East Anglia and Lincolnshire is made of weak sedimentary clays, which form extensive flat agricultural plains.
- Lowland 'vales' (wide valleys) occur throughout southern England in between the chalk and limestone escarpments.
- The Central Lowlands of Scotland is a rift valley formed by the downward slippage of rock in between two faults, the Highland Boundary Fault to the north and the Southern Uplands Fault to the south.
- Numerous other lowlands correspond with river valleys, such as the Severn, the Thames and the Trent.

Where are the UK's river systems?

REVISED

The UK has extensive river systems. Most rivers have their source in the mountains, often radiating outwards as they head to the coast. Notice, for example, that major Scottish rivers such as the Spey, Tay (at 187 kilometres in length, Scotland's longest river), and Forth all radiate out from the Grampian Mountains.

Several major English rivers flow eastwards from their source in the Pennine Hills, such as the Tyne, Tees and Ouse.

The longest river in the UK is the River Severn (352 kilometres), which has its source in the Cambrian Mountains in Wales, with the River Thames a close second at 344 kilometres.

> **Exam practice**
>
> Describe the distribution of uplands and lowlands in the UK. (4 marks)
>
> ONLINE

> **Now test yourself**
>
> Find a relief map of the UK and use it to describe the relief of Scotland.
>
> TESTED

10 Coastal landscapes in the UK

10.1 Coastal processes

What are the main wave types and their characteristics?

REVISED

Waves are formed by the wind blowing over the sea. Friction with the water's surface causes ripples to form, which can then develop into waves. The energy of the waves is determined by:

- the strength of the wind
- the duration of the wind
- the distance of open water over which the wind blows – this is called the fetch.

In the open water, waves have a circular (orbital) motion. As the waves approach the shore, this orbital motion is interrupted by the shallowing seafloor, causing the waves to rise up and eventually break on the **beach** (Figure 10.1).

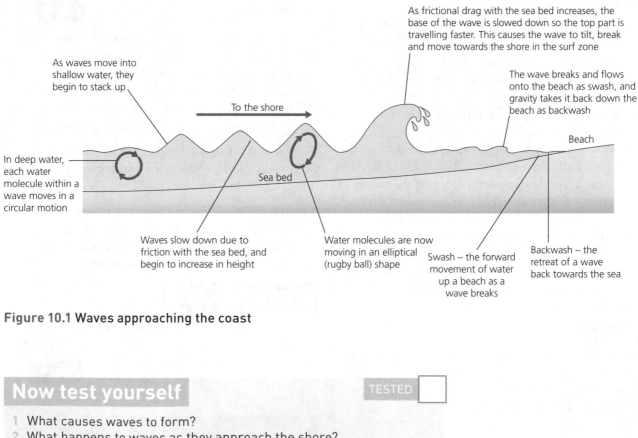

As waves move into shallow water, they begin to stack up

As frictional drag with the sea bed increases, the base of the wave is slowed down so the top part is travelling faster. This causes the wave to tilt, break and move towards the shore in the surf zone

The wave breaks and flows onto the beach as swash, and gravity takes it back down the beach as backwash

To the shore

Beach

In deep water, each water molecule within a wave moves in a circular motion

Sea bed

Waves slow down due to friction with the sea bed, and begin to increase in height

Water molecules are now moving in an elliptical (rugby ball) shape

Swash – the forward movement of water up a beach as a wave breaks

Backwash – the retreat of a wave back towards the sea

Figure 10.1 Waves approaching the coast

Now test yourself

TESTED

1 What causes waves to form?
2 What happens to waves as they approach the shore?

Constructive and destructive waves

It is possible to identify two types of wave breaking at the coast: constructive (Figure 10.2) and destructive (Figure 10.3).

● Constructive waves – these are low waves with long wavelengths formed by distant storms. On breaking, they surge up the beach with a strong swash. Water soaks into the beach, resulting in a weak backwash. Over time, they build up a beach, hence the term 'constructive'.

● Destructive waves – these are higher waves with shorter wavelengths formed by local storms. As they crash down (plunge) onto the beach, there is little forward movement of water (swash) but powerful backwash. Over time, this results in the lower beach being eroded.

Wave types vary throughout the year, with constructive waves being more common in the summer and destructive being more common in the winter, when frequent storms approach the UK. This explains why beaches vary in their profiles and material during the course of the year.

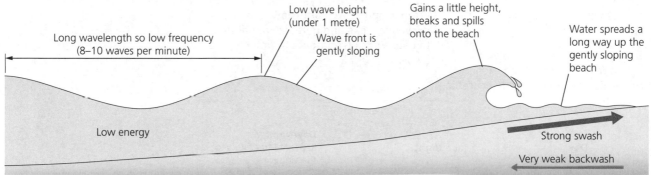

The strong swash transports sand and shingle up the beach to construct it

Figure 10.2 Constructive waves

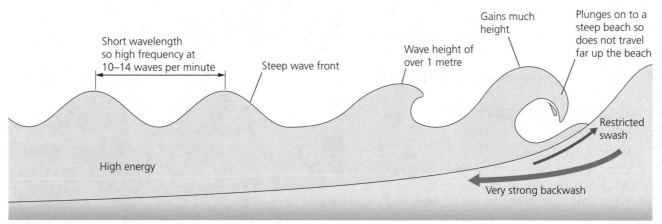

The relatively stronger backwash drags beach material out to sea

Figure 10.3 Destructive waves

Now test yourself

 TESTED ☐

1 Name two characteristics of constructive waves.
2 Name two characteristics of destructive waves.

What are the main coastal processes?

Weathering

Weathering involves the decomposition or disintegration of rock in its original place at or close to the ground surface. There are two main types of weathering: **chemical weathering** and **mechanical (physical) weathering**. A third type, biological weathering, involves living organisms such as nesting birds, burrowing rabbits and plant roots.

Chemical weathering	Mechanical weathering
Carbonation – carbon dioxide dissolved in rainwater forms a weak carbonic acid. This reacts with calcium carbonate (limestone and chalk) to form calcium bicarbonate, which is soluble and can be carried away by water.	Freeze-thaw – repeated cycles of freezing and thawing causing water trapped in rocks to expand/contract, eventually causing rock fragments to break away (Figure 10.4).
Hydrolysis – acidic rainwater reacts with minerals in granite, causing it to crumble.	Salt weathering – crystals of salt, often evaporated from seawater, grow in cracks and holes, expanding to cause rock fragments to flake away.
Oxidation – oxygen dissolved in water reacts with iron-rich minerals, causing rocks to crumble.	

Day

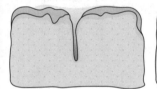

Night

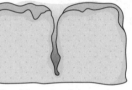

Day/Night: Repeated freezing and thawing

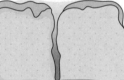

Water seeps into cracks and fractures in the rock.

When the water freezes, it expands about 9% in volume, which wedges apart the rock.

With repeated freeze-thaw cycles, the rock breaks into pieces.

Figure 10.4 The freeze-thaw weathering cycle

Now test yourself

1 What is carbonation?
2 What type of weathering does carbonation fall under?
3 What is freeze-thaw weathering?
4 What type of weathering does freeze-thaw fall under?

Mass movement

Mass movement is active at the coast, particularly where **cliffs** are undercut by the sea, making them unstable. It includes **sliding** and **slumping** as well as falls (rockfalls) and flows (mudflows). Common forms of mass movement at the coast include:

- rockfall – individual fragments or chunks of rock falling off a cliff face, often resulting from freeze-thaw weathering (Figure 10.5)
- landslide – sliding of blocks of rock moving rapidly downslope along a linear shear-plane, usually lubricated by water (Figure 10.6)
- mudflow – saturated material (usually clay) flowing downhill, which may involve elements of sliding or slumping as well as flow
- rotational slip/slump – slumping of loose material, often along a curved surface lubricated by water (Figure 10.7).

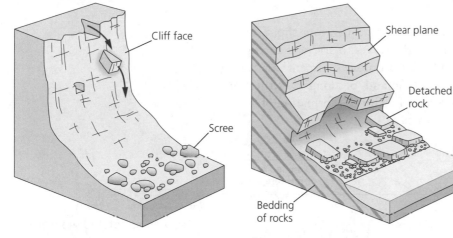

Figure 10.5 Rockfall

Figure 10.6 Landslide

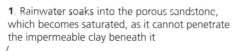

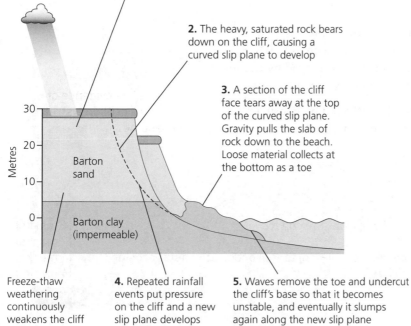

1. Rainwater soaks into the porous sandstone, which becomes saturated, as it cannot penetrate the impermeable clay beneath it

2. The heavy, saturated rock bears down on the cliff, causing a curved slip plane to develop

3. A section of the cliff face tears away at the top of the curved slip plane. Gravity pulls the slab of rock down to the beach. Loose material collects at the bottom as a toe

Freeze-thaw weathering continuously weakens the cliff

4. Repeated rainfall events put pressure on the cliff and a new slip plane develops

5. Waves remove the toe and undercut the cliff's base so that it becomes unstable, and eventually it slumps again along the new slip plane

Figure 10.7 Slumping at Barton-on-Sea, Hampshire

Revision activity

Make a copy of Figure 10.6 and add labels to describe the causes and characteristics of sliding.

Now test yourself

Use Figure 10.7 to describe the causes and characteristics of slumping.

TESTED ☐

Coastal erosion

Coastal erosion involves the removal of material and sculpting of landforms. The main processes of erosion are **hydraulic power**, **abrasion/corrasion** and **attrition**. An additional process – solution – involves the dissolving of soluble rocks, such as chalk and limestone. Coastal erosion processes work together to create landforms such as cliffs and **wave-cut platforms**.

Coastal transportation

Coastal **transportation** involves the following processes:
- Traction – large particles rolling along the seabed
- Saltation – a bouncing or hopping motion by pebbles too heavy to be suspended
- Suspension – particles suspended within the water
- Solution – chemicals dissolved in the water.

The movement of sediment along the coastline is called **longshore drift**. It occurs when the wind direction drives the waves to arrive at an angle to the coast (Figure 10.8). Swash and backwash move sediment in a zigzag way along the beach to pile up against a headland, or alongside structures such as **groynes**. Longshore drift is responsible for the formation of coastal landforms including spits and beaches.

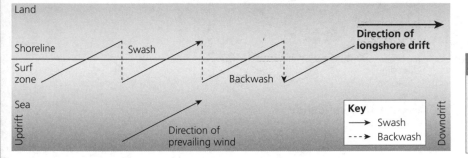

Figure 10.8 Longshore drift

Coastal deposition

Coastal **deposition** is where sediment carried by the sea is dropped and left behind. It typically occurs in areas of low wave energy, where velocity is reduced and sediment can no longer be transported by the sea. Deposition is common in bays or in areas sheltered by bars and spits.

Coastal deposition also occurs on wide beaches affected by constructive waves, close to rapidly eroding cliffs and on the updrift side of engineering structures such as groynes.

Revision activity

Draw a simple diagram to show the four processes of coastal transportation.

Revision activity

Draw an annotated diagram to describe the process of longshore drift, without looking at the one here.

Now test yourself

Describe the evidence of longshore drift on a stretch of coastline.

TESTED ☐

Exam practice

1 What factors affect the characteristics of waves at the coast? (4 marks)
2 What is mechanical weathering? (2 marks)
3 Under what conditions will hydraulic power be an important process of coastal erosion? (4 marks)
4 Why is sediment deposited in coastal areas? (6 marks)

ONLINE ☐

10.2 Coastal landforms

How does geology affect coastal landforms?

Rock type and geological structure

Geology can affect coastal landscapes and landforms in two ways:

- Rock type – rocks vary in their strength and resistance to erosion. Hard, resistant rocks, such as granite and limestone, withstand erosion to form tall cliffs or coastal headlands. Weaker, soft rocks, such as clays and sands, are more easily eroded and tend to form low cliffs or bays.
- Geological structure – this is to do with the arrangement of rocks and whether the layers (beds) have been folded or faulted. Folded rocks exposed at the coast can affect the shape of the cliff profile. If the rocks are horizontal, a stepped profile is likely, whereas if the rocks are dipping vertically, a steep cliff face is more likely. Faults can form lines of weakness at the coast to be exploited by processes of erosion. Figure 10.9 shows the effects of geology at the coast.

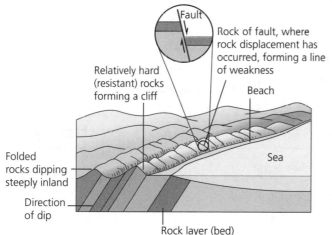
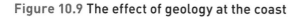

Figure 10.9 The effect of geology at the coast

The effect of geology on the Dorset coast

Figure 10.10 shows the effect of rock type on the Dorset coast. Alternate hard and soft bands of rock are exposed along the east-facing coastline.

- The tougher bands of chalk and limestone are more resistant to erosion and stick out into the sea to form headlands.
- The weaker Wealden clay has been more easily eroded to form Swanage Bay.
- This stretch of coastline, where rocks reach the coast at right angles creating headlands and bays, is called a discordant coastline.

In contrast, along the southern coast, the rocks are now parallel to the coast, forming a more or less straight section of cliffed coast, particularly where the resistant Portland limestone is exposed. This is called a concordant coastline.

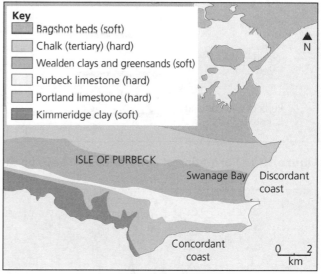

Figure 10.10 The effect of rock type on the Dorset coast

Now test yourself

1 How does rock type affect the development of landforms on the Dorset coast?
2 What is the difference between a concordant and discordant coastline?

What are the main landforms of coastal erosion?

Headlands and bays

Headlands and bays are characteristic features of a discordant coastline where rocks of different hardness (resistance to erosion) are exposed at the coast. Look at Figure 10.11. Notice that coastal erosion processes erode away the weaker rocks more readily than the harder rocks, to form a sequence of alternating headlands and bays.

As the coastline becomes more indented the headlands start to interfere with the incoming waves, causing them to be bent or refracted (Figure 10.12). This results in energy gradually becoming more concentrated at the headlands and less concentrated in the bays. In this way, the bays cease to become eroded and the coastline reaches a state of balance.

- Headlands – jutting out into the sea, headlands (e.g. The Foreland) are more exposed to high-energy waves and are likely to exhibit landforms of erosion such as cliffs, wave-cut platforms and stacks.
- Bays – sheltered by the headlands, bays (e.g. Swanage Bay) tend to experience low-energy waves, resulting in deposition and the formation of beaches.

> **Revision activity**
>
> Make up your own diagram to show the formation of headlands and bays. Draw wave fronts to show wave refraction. Add labels to describe what is happening.

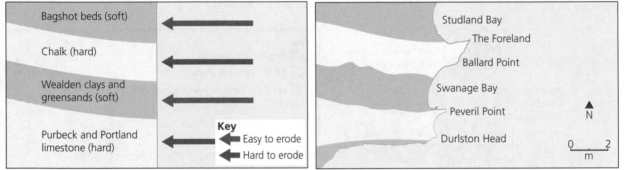

Figure 10.11 Formation of headlands and bays on the Dorset coast

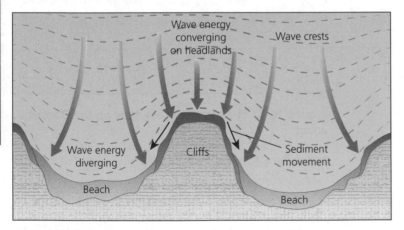

Figure 10.12 Wave refraction

Cliffs and wave-cut platforms

Cliffs and wave-cut platforms are common landforms at the coast, particularly where hard rocks such as granite, limestone and chalk reach the coast. Figure 10.13 describes the formation of these landforms.

- When waves break against the foot of a cliff, they erode a wave-cut notch through the processes of hydraulic power and abrasion (corrosion).
- This undercuts the cliff face, which eventually collapses through mass movement (e.g. rockfall or landslide).
- The cliff profile is steepened and the cliff line gradually retreats inland, leaving behind an extensive rocky wave-cut platform, which becomes smoothed by pebbles constantly grinding over it (abrasion).

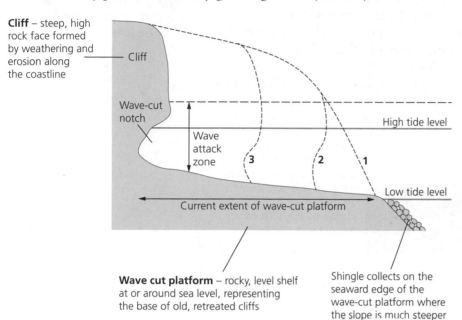

Exam tip

Use annotated diagrams to support your answer when describing the formation of coastal landforms.

Cliff – steep, high rock face formed by weathering and erosion along the coastline

Cliff

Wave-cut notch

Wave attack zone

High tide level

3   2   1

Low tide level

Current extent of wave-cut platform

Wave cut platform – rocky, level shelf at or around sea level, representing the base of old, retreated cliffs

Shingle collects on the seaward edge of the wave-cut platform where the slope is much steeper

Figure 10.13 Formation of cliffs and wave-cut platforms

Caves, arches and stacks

Caves, arches and **stacks** are commonly formed at headlands, where relatively tough rock juts out into the sea. It is the gradual erosion of this headland that leads to the formation of these distinctive coastal landforms (Figure 10.14).

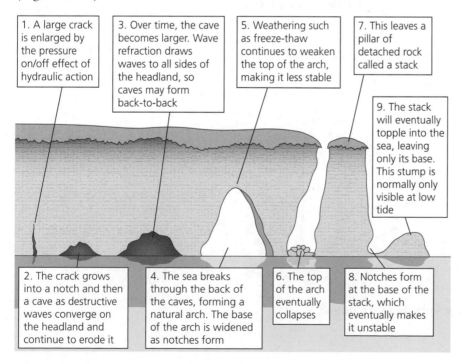

1. A large crack is enlarged by the pressure on/off effect of hydraulic action

3. Over time, the cave becomes larger. Wave refraction draws waves to all sides of the headland, so caves may form back-to-back

5. Weathering such as freeze-thaw continues to weaken the top of the arch, making it less stable

7. This leaves a pillar of detached rock called a stack

9. The stack will eventually topple into the sea, leaving only its base. This stump is normally only visible at low tide

2. The crack grows into a notch and then a cave as destructive waves converge on the headland and continue to erode it

4. The sea breaks through the back of the caves, forming a natural arch. The base of the arch is widened as notches form

6. The top of the arch eventually collapses

8. Notches form at the base of the stack, which eventually makes it unstable

Figure 10.14 Formation of caves, arches and stacks

Revision activity

Draw a series of annotated diagrams to show the formation of a stack.

What are the main landforms of coastal deposition?

REVISED

Beaches

A beach is a depositional landform made of sand or pebbles (shingle) found at the coast. Beaches may exhibit a range of small landforms such as ridges, called berms (formed by waves just above the high tide line, for example), ripples, or shallow water-filled depressions called runnels.

- Sandy beaches – most commonly formed in sheltered bays, associated with relatively low-energy constructive waves. Flat and extensive beaches are often backed by sand dunes.
- Pebble beaches – commonly associated with higher energy coastlines where destructive waves remove finer sand, leaving behind coarser pebbles. Pebble beaches tend to be steep and narrow with distinctive high tide berms.

Now test yourself

Suggest reasons why some stretches of coastline have sandy beaches.

TESTED

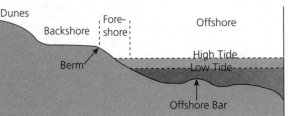

Offshore – lowest level of wave action to low tide

Foreshore – from low tide to just above high tide, usually marked by a berm

Backshore – only affected by storm waves so mostly dry

Figure 10.15 Beach profile

Sand dunes

Sand dunes form at the back of broad sandy beaches. Sand, deposited by the sea on the beach, dries out at low tide. It is then picked up and carried inland by the wind. Sand-dune formation involves a number of stages (Figure 10.16):

- Sand accumulates against obstacles on the beach such as driftwood.
- Sand builds up to form small mobile embryo dunes.
- Hardy plants such as marram grass colonise and help to stabilise the dunes, forming distinctive tall foredunes.
- Dead organic matter adds nutrients to the sand, enabling a wider range of plants to grow on the stable yellow and grey dunes, and biodiversity increases.
- In places, dune slacks may form where wind hollows out a depression exposing the water table.
- Trees and shrubs grow on the oldest sand dunes furthest away from the sea.

As Figure 10.16 shows, these stages form a distinctive sequence of sand dunes.

Revision activity

Make a copy of Figure 10.16 and add annotations to explain the changing characteristics with distance away from the sea.

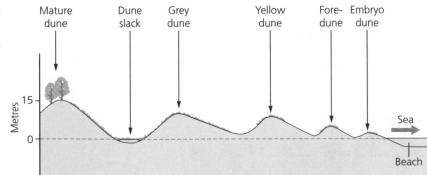

Figure 10.16 Changes in sand dunes with distance inland

Spits

A **spit** is a sand or shingle (pebble) ridge most commonly formed by longshore drift operating along a stretch of coastline (Figure 10.17). It is rather like a narrow beach extending out from the coast into the sea. At its end, where it is more exposed to variations in wind and waves, it tends to curve to form a hook or recurved tip. Some spits have sand dunes on them.

Exam tip

Remember that a spit is land, so it lies above the high tide line. On OS maps, use the high tide line to trace the edges of a spit.

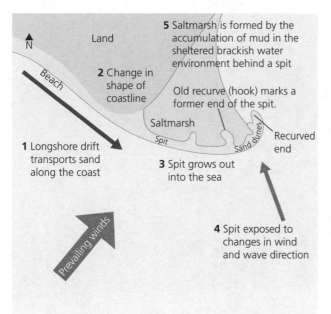

Figure 10.17 Formation of a spit

Bars

A bay **bar** is a ridge of sand or shingle that extends across a bay, creating a lagoon. It is formed by longshore drift transporting sediment from one side of a bay to the other (Figure 10.18).

A bar can also form offshore. When sand or shingle is eroded and dragged back from a beach by destructive waves, it may be deposited as a ridge a short distance out to sea. At low tide, this ridge may be exposed as a bar. At high tide, it may be marked by breakers as waves are forced to break early by the shallow water.

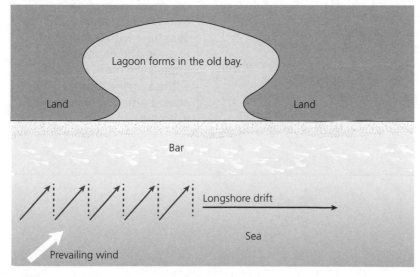

Figure 10.18 The formation of a bay bar

Now test yourself

Describe the formation of bars.

TESTED

Coastal landforms on the Dorset coast, near Swanage

The Dorset coast near Swanage exhibits several landforms of coastal erosion and deposition (Figure 10.19).

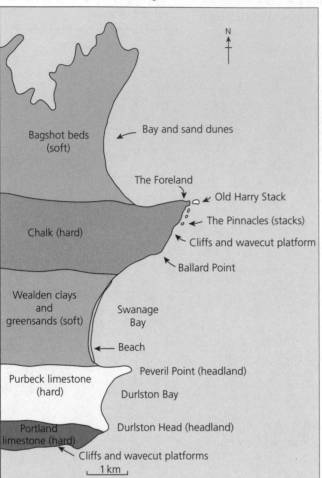

Figure 10.19 **Coastal landforms on the Dorset coast**

Exam practice

1 Describe how rock type can influence coastal landforms. (4 marks)
2 Explain the formation of cliffs and wave-cut platforms. (6 marks)
3 Explain how longshore drift leads to the formation of spits and bars. (6 marks)
4 With reference to an example of a section of coastline, describe the major coastal landforms of deposition. (6 marks)

ONLINE

You need to know an example of a section of coastline with features of erosion and deposition, so make sure that you learn some specific landforms and place names for your chosen example.

10.3 Coastal management strategies

What are the options for coastal management?

The coastal zone is where physical (natural) forces and human activities interact. It needs to be carefully managed to protect people and property while conserving the natural environment.

Management schemes have economic, social and environmental costs which have to be compared to the benefits offered by protection. Usually it is the economic costs and benefits that are most important in deciding whether a management scheme will go ahead.

There are four management options:
1 Do nothing – let nature take its course. This applies to much of the UK coastline where there is no serious threat or where the value of the land is low.
2 Hold the line – use coastal defences to retain the line of the current coastline, preventing recession or flooding inland.
3 Advance the line – use coastal defences and schemes aimed at extending the land towards the sea to provide greater protection for coastal developments.
4 **Managed retreat** – allow the current coastline to retreat inland in a controlled way by allowing land to become deliberately flooded.

Two types of coastal engineering can be adopted to protect the coast: **hard engineering** and **soft engineering**.

What are the main hard engineering strategies?

Strategy	Your sketch/photo	Costs	Benefits
Sea wall – commonly a concrete structure placed at the top of a beach or foot of a cliff to act as a physical barrier to the sea, preventing erosion or flooding.		<ul><li>£5,000–£10,000 per metre.</li><li>Very expensive, with high maintenance costs.</li><li>Can look unattractive and obtrusive; very artificial in the natural landscape.</li><li>Interference with waves can lead to local scouring (undercutting of the sea wall) or refraction, leading to high-energy waves breaking elsewhere along the coast.</li></ul>	<ul><li>Usually very successful as a direct barrier to the sea.</li><li>Opportunities for developing the top of the sea wall as an amenity (walking, street stalls, etc.).</li></ul>
Rock armour – large, extremely tough boulders placed at the foot of a cliff or against a sea wall, forcing waves to break early, reducing their energy and protecting the coast from their full force.		<ul><li>£200,000 per 100 metres.</li><li>Boulders usually different rock type to the local area (much of it is imported from Norway).</li><li>Can be ugly and obtrusive.</li><li>Potentially dangerous to the public.</li></ul>	<ul><li>Relatively cheap to construct and maintain – boulders arrive by barge.</li><li>Some amenity use, e.g. fishing.</li></ul>

Strategy	Your sketch/photo	Costs	Benefits
Gabions – wire cages filled with rocks commonly built up against a cliff to add support and reduce erosion. Being permeable, they improve cliff drainage.		• Up to £50,000 per 100 metres. • Metal cages with rocks are not especially attractive. • Cages may break apart in storms, becoming dangerous, unattractive and ineffective. • After about ten years, their functionality is reduced as they start to rust.	• Very flexible construction options – like Lego building blocks! • Relatively cheap and quick to construct and maintain. • Quickly 'green' over as plants (e.g. brambles) colonise the area.
Groynes – timber or rock structures protruding into the sea at right angles to the coast. Sediment is trapped between the groynes, broadening the beach and affording greater protection to the coast by absorbing wave energy.		• Timber groynes can cost up to £150,000 each – usually constructed at 200-metre intervals. • Look artificial and some consider groynes to be an eyesore. • By trapping sediment carried by longshore drift, they starve beaches further down-drift, often increasing the risk of erosion and flooding. • Several groynes are often needed along a stretch of coast, increasing the cost. • Regular maintenance is needed, particularly after storms or when they become overtopped by deposited sediment.	• Very effective in trapping sediment transported by longshore drift. • Wider beach is a good amenity, providing opportunities for tourism.

What are the main soft engineering strategies?

Strategy	Your sketch/photo	Costs	Benefits
Beach nourishment and **beach reprofiling** – the addition (feeding) of sand or pebbles to a beach to increase its height and/or width, providing protection from erosion or flooding by absorbing wave energy. Bulldozers can reprofile the beach to create high ridges.		• Up to £500,000 per 100 metres (but can vary hugely depending on transport costs and quantity). • Constant maintenance and reprofiling is required, particularly after winter storms. • People may be prevented from using the beach for several weeks during maintenance.	• Usually looks very natural and can improve the attractiveness of a stretch of coast. • Creates a useful amenity for tourism. • Relatively cheap and easy to maintain; barges bring the beach material onshore.
Dune regeneration – plants such as marram grass can be sown to stabilise sand dunes and encourage them to develop, acting as a natural buffer to the sea. Fences can be used to protect dunes from human use.		• £400–£2,000 per 100 metres. • Time-consuming to plant the grass and maintain the area, keeping people off the newly planted vegetation. • Can be easily damaged by storms.	• Considered natural by most people. • Can produce an attractive amenity for tourists – walking, picnics. • May increase biodiversity, providing a greater range of natural habitats for plants, animals and birds.

Now test yourself

TESTED

1 What is the difference between hard and soft engineering?
2 How do groynes help to protect a stretch of coast from erosion and/or flooding?
3 Compare the costs and benefits of constructing a sea wall with using beach nourishment.

Revision activity

Draw labelled sketches, or find appropriate photos, for the tables above to illustrate the hard and soft engineering strategies described.

What is managed retreat (coastal realignment)?

In the past, coastal management involved protecting coastal developments and agricultural land at almost any cost. With rising sea levels and increased storminess associated with climate change, this policy has been reviewed, particularly in low-lying areas where the land values are low and the cost of sea defences high.

Managed retreat – or coastal realignment – deliberately allows the sea to erode or flood an area in a controlled way. For flooding, managed retreat works as follows:

- A cost-benefit analysis is conducted prior to a decision being made.
- If costs outweigh the benefits, an area of land is identified for deliberate flooding. Land owners will be compensated.
- Low earth embankments may be constructed inland to protect higher value land, property or roads. Footpaths will be re-routed.
- Old sea defences are deliberately breached to allow seawater to encroach over the land.
- Gradually the flooded land turns into saltmarsh, establishing an important new wetland environment. As the saltmarsh builds up, it creates a natural, sustainable buffer to the sea.

Costs	Benefits
<ul><li>Costs will depend on the scale of the scheme, usually several million pounds.</li><li>Some low-value land will be lost to the sea – this could be farmland.</li><li>Local people may lose land, have to be relocated or lose access (footpaths).</li><li>Some ecosystems may be affected by flooding.</li></ul>	<ul><li>Long-term, sustainable solution with very low maintenance costs.</li><li>Creation of a natural buffer to the sea (saltmarsh), increasing protection of inland areas.</li><li>Creation of a new saltmarsh ecosystem.</li><li>Increased tourism – birders, walkers, etc.</li><li>More attractive than using other forms of coastal defence.</li></ul>

Now test yourself

What is managed retreat and why has it become a popular option in recent years?

What are some examples of coastal management?

	Lyme Regis	Medmerry
Reasons for management	• Rapidly eroding weak cliffs (limestone and shales) along this section of the World Heritage Jurassic Coast. • Coast prone to landslips, threatening properties. • Seafront properties and businesses in Lyme Regis threatened by high waves during storms. • Lack of beach amenity, with waves having eroded away much of the beach – tourist numbers were down, affecting businesses.	• Cost of maintaining the low sea wall (pebble ridge): £200,000 a year. • Breaching of the pebble ridge was becoming more frequent (2008: damage of £5 million). • Projected sea level rise means breaches were more likely and maintenance costs would increase. • Much of the land adjacent to the coast was relatively low-value farmland. • Properties in nearby Selsey, a water treatment works, roads and holiday homes/caravan parks were under threat.
Management strategy	• Lyme Regis Environmental Improvement Scheme (1990s–2015) has involved a mixture of hard and soft engineering. • Hard engineering – new sea walls and promenades, rock armour to protect the harbour wall (the Cobb) and at the eastern end of the town. • Soft engineering – widespread beach nourishment (pebbles from English Channel, sand from France). • Nailing and drainage of the cliffs, together with some reprofiling to stabilise and improve drainage. • Total cost: £40 million plus.	• Environment Agency decided to follow the strategy of managed retreat, deliberately breaching the old sea wall (in 2013) to allow farmland to be flooded. • Compensation was paid to local landowners and footpaths redirected. • A 2-kilometre embankment was constructed around the perimeter of the area to be flooded, protecting roads, farmland and caravan parks. • Rock armour was placed at the seaward edges of the embankment to provide extra protection. • Total cost: £28 million.
Effects	• Huge improvement to the attractiveness of the seafront and amenity (sandy beach, wide promenade). • Effective protection from powerful storms. • Increased trade for businesses as tourist numbers have increased. • Increased stability of slopes and peace of mind for homeowners.	• Properties in Selsey and caravan park owners are better protected. • New footpaths, cycle tracks and bridleways have opened. • The newly forming saltmarsh will create an effective buffer to the sea and will attract wildlife, particularly waterfowl and overwintering birds.
Conflicts	• Increased tourism has led to some conflicts with local people (traffic congestion, parking, increased prices). • Some people think the schemes have spoilt the natural landscape. • Reduced erosion of the cliffs results in fewer important fossils being eroded out of the cliffs. • Coastal processes may be disrupted because of the new sea walls.	• Some conflicts with local people concerned about loss of habitats, footpath access, etc. • Some people from outside the area felt that it was an expensive scheme for an area with low population. • Farmland that used to grow rapeseed has been lost.

Exam practice

1 What is meant by soft engineering? (2 marks)
2 How can soft engineering provide coastal protection as well as improving the natural environment? (6 marks)
3 Using an example, evaluate the success of a coastal management scheme in the UK. (6 marks)

ONLINE

Exam tip

When evaluating, make sure you consider equally the pros (advantages) and cons (disadvantages). Be prepared to make a supported judgement.

Revision activity

Draw a simple sketch map of your chosen management scheme, showing the strategies that have been adopted.

11 River landscapes in the UK

11.1 Rivers and river valleys

River valleys – what are long and cross profiles?

Long profiles

The **long profile** of a river takes the form of a concave curve. The steepest gradient is in the uplands near the source of the river. Further downstream, the gradient is reduced as the river flows from uplands to lowlands and on to the sea.

There may be slight irregularities in the long profile of a river due to the geology. For example, bands of tough, resistant rock may form 'steps' in the profile marked by **waterfalls**.

Cross profiles

The **cross profile** of a river and its valley changes with distance downsteam. Figure 11.1 shows the changes in the river channel cross profile. Notice that the river channel becomes wider and deeper with distance downstream. As tributaries join the river, an increasing volume of water carves an ever-larger channel.

The river valley profile also changes with distance downstream (Figure 11.2).

- Upper course – narrow and steep-sided, due to **vertical erosion** by the river combined with weathering and mass movement of the valley slopes.
- Middle course – wider valley floor formed by lateral (sideways) erosion of a meandering river, with gentler valley side-slopes.
- Lower course – almost flat valley floor comprising wide floodplain and elaborately meandering river. Much of the valley floor is sediment dumped by the river, accounting for its flatness.

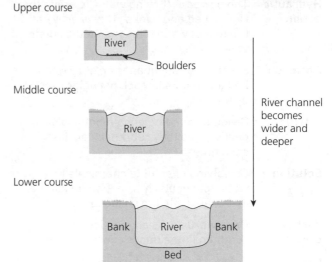

Figure 11.1 River channel cross-profile changes with distance downstream

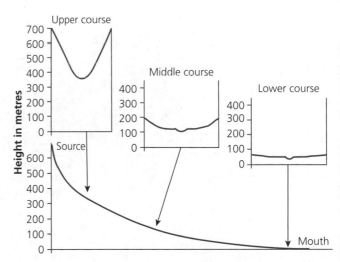

Figure 11.2 River valley long-profile changes with distance downstream

1 Describe the long profile of a river from source to mouth.
2 Why does a river channel get deeper and wider with distance downstream?
3 Describe and account for the shape of a river valley in its upper course.

What are fluvial processes?

There are several types of **fluvial (river) processes** you will need to revise.

Erosion

This involves the picking up and removal of loose material or sculpting landforms.

Process	Description
Hydraulic action	The power of flowing water to erode the river's bed and banks. This is active at the foot of waterfalls and on the outside bends of meanders.
Abrasion	Scraping or sandpapering of a river's bed and banks by rock particles carried by the river.
Attrition	Gradual rounding and smoothing of rock particles as they rub/knock against each other.
Solution	Dissolving of soluble chemicals in water, particularly affecting limestone and chalk (calcium carbonate).
Vertical erosion	Downwards erosion, particularly common in the upper course of a river.
Lateral erosion	Sideways erosion, particularly common in the middle and lower courses of a river.

Transportation

This involves the transfer of sediment by a river.

Process	Description
Traction	Large particles rolled along the river bed by the force of the water.
Saltation	A bouncing or hopping motion by pebbles too heavy to be suspended.
Suspension	Particles suspended within the water.
Solution	Chemicals dissolved in the water.

Deposition

This involves the dropping of sediment that has been transported by a river.

River sediment is deposited in low flow conditions when the velocity of the river can no longer carry the sediment load. This can occur anywhere along a river's long profile where velocity falls.

Figure 11.3 shows the main fluvial environments where deposition occurs.

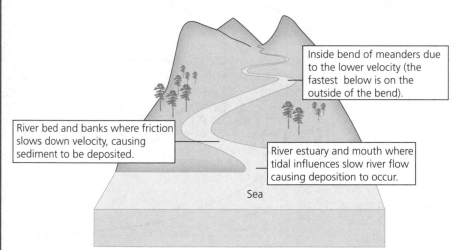

Inside bend of meanders due to the lower velocity (the fastest below is on the outside of the bend).

River bed and banks where friction slows down velocity, causing sediment to be deposited.

River estuary and mouth where tidal influences slow river flow causing deposition to occur.

Sea

Figure 11.3 Deposition environments in a river

Revision activity

Use simple diagrams to describe the processes of river erosion.

Exam tip

In answering Question 1, make sure you focus on the **river valley**. Do not write about changes in the **river channel**.

Exam practice

1 Describe the changes in a river valley with distance downstream. (4 marks)
2 Explain how a river erodes its channel. (4 marks)
3 Under what conditions does a river deposit its sediment? (4 marks)

ONLINE

Exam tip

Remember to name specific fluvial processes of erosion and transportation as these are specific geographical terms and will gain you credit.

11.2 River landforms

What landforms result from erosion?

Interlocking spurs

Interlocking spurs are commonly found in a river's upper course. As small streams and rivers tumble down mountainsides, they are forced to flow around 'fingers' of land that jut out into the river valley. It is these 'fingers' of land that are the interlocking spurs (Figure 11.4).

Vertical rather than **lateral erosion** dominates in the upper course of a river. This deepens the valley but does not erode sideways into the interlocking spurs. This explains why the river is forced to flow around them.

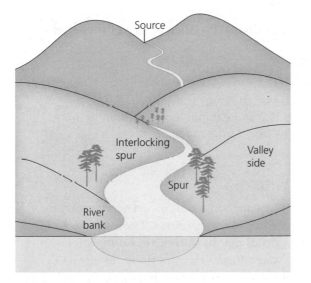

Figure 11.4 Interlocking spurs

Waterfalls

A waterfall is a 'step' in the long profile of a valley. Typically, fast-flowing water plummets over a vertical cliff – often a considerable drop – into a deep plunge pool below (Figure 11.5). Hydraulic action and abrasion are mainly responsible for eroding the plunge pool, which can be several metres deep in the centre.

Waterfalls most commonly form when a river flows over a hard, resistant band of rock. Unable to erode the tougher rock, a 'step' is formed in the long profile of the river. It is at this point where the waterfall forms.

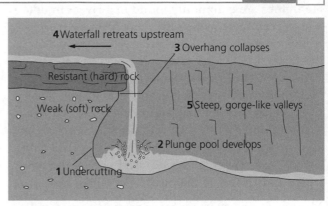

Figure 11.5 The formation of a waterfall

Now test yourself

Describe the characteristics of a waterfall.

Gorges

A **gorge** is commonly formed by the upstream retreat of a waterfall (Figure 11.6).

- Erosion (primarily hydraulic action and abrasion) undercuts the hard rock, forming the waterfall and creating an overhang.
- Eventually this overhang collapses into the plunge pool, causing the waterfall to retreat upstream.
- Over many thousands of years, this process of undercutting and collapse continues and a gorge is formed immediately downstream.

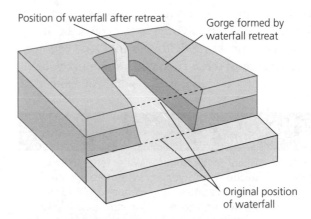

Figure 11.6 The formation of a gorge

Revision activity

Draw a series of labelled diagrams to describe the formation of a gorge.

What landforms result from erosion and deposition?

Meanders

Meanders are commonly found in a river's middle and lower course, where they can form extensive and elaborate bends. The fastest velocity – called the thalweg – swings around the outside bend of a meander, eroding the banks to form a river cliff. Here, the water is deep. On the inside bend, where the velocity is lower, deposition occurs, forming a slip-off slope. In this way, the meander develops an asymmetrical cross profile (Figure 11.7). Over time, lateral erosion on the outside bend widens the river valley and creates an extensive, flat floodplain.

Meandering rivers are most commonly associated with the following environmental conditions:
- Gentle gradients.
- Relatively fine sediments.
- Steady precipitation throughout the year.

This explains why they are by far the most common river pattern in the UK and, indeed, throughout the world.

A common feature of meandering rivers is an alternating pattern of shallows (riffles) and deeps (pools). These features are associated with a complex corkscrew-like motion of velocity within the river called helicoidal flow.
- Riffles – these shallow areas are associated with the straighter sections of rivers in between meanders. They usually have rocky beds and turbulent flow due to friction with the river bed.
- Pools – these deeper areas are associated with the meander bends. They usually have finer sediment and less turbulent flow due to the smoother river bed.

The relationship between velocity, channel size and friction is complex, resulting in varying rates of flow between riffles and pools during the course of the year. When conducting fieldwork in rivers, you need to take account of these factors in explaining your results. You may not always find what you expect!

Rate of flow	Riffle	Pool
High flow (winter)	Greater friction in the shallower riffles results in slower, more turbulent flow.	Water tends to flow faster through the deeper pools due to a reduction in friction with the bed and banks.
Low flow (summer)	On entering a riffle, the reduction in channel size often results in slightly faster flow.	The reduced volume of water tends to slow down on entering a deep pool, where the channel is larger.

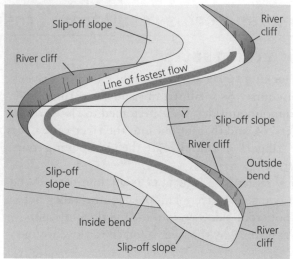

Cross profile

Figure 11.7 Characteristics of a river meander

Draw a sketch cross profile from X–Y on Figure 11.7 and explain the formation of the features of erosion and deposition.

Oxbow lakes

Oxbow lakes are commonly associated with rivers in their middle of lower course. They represent old meander bends that have been cut off by faster-flowing water during times of flood (Figure 11.8). Gradually, without water flowing through them, they will become silted and colonised by marshy vegetation.

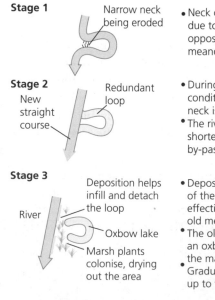

Stage 1 Narrow neck being eroded
- Neck of meander narrows due to lateral erosion on opposite sides of the meander bend

Stage 2 New straight course / Redundant loop
- During high flow (flood) conditions, the meander neck is broken through
- The river now adopts the shorter (steeper) route, by-passing the old meander

Stage 3 River / Deposition helps infill and detach the loop / Oxbow lake / Marsh plants colonise, drying out the area
- Deposition occurs at the edges of the new straight section, effectively cutting off the old meander
- The old meander now forms an oxbow lake, separated from the main river
- Gradually the oxbow lake silts up to form marshland

Figure 11.8 Formation of an oxbow lake

What landforms result from deposition?

Levées

Levées are raised river banks commonly found in the lower course of rivers. They are formed during flood conditions, when water flows over the river banks onto the surrounding floodplain (Figure 11.9).

● As water overtops the river banks, there is a sudden localised reduction in velocity of the water that had previously been flowing very fast along the river channel.

● This causes sediment in suspension to be deposited at the river bank.

● Coarse sediment (heavier) is deposited first and this traps the finer sediment.

● With each successive flood, the deposited sediment raises the river banks by as much as a few metres.

Levées can be created artificially by people to contain water within a river channel to reduce the threat of flooding. In the USA, the term 'levée' is most commonly used for the artificial form.

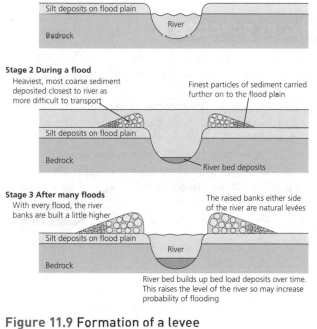

Stage 1 Before levée
Silt deposits on flood plain
River
Bedrock

Stage 2 During a flood
Heaviest, most coarse sediment deposited closest to river as more difficult to transport
Finest particles of sediment carried further on to the flood plain
Silt deposits on flood plain
Bedrock
River bed deposits

Stage 3 After many floods
With every flood, the river banks are built a little higher
The raised banks either side of the river are natural levées
Silt deposits on flood plain
River
Bedrock
River bed builds up bed load deposits over time. This raises the level of the river so may increase probability of flooding

Figure 11.9 Formation of a levee

Floodplains

Floodplains are associated with rivers in their middle and lower course. They are extensive, flat areas of land mostly covered by grass. There may be some marshy areas close to the river and also oxbow lakes (Figure 11.10). As the name implies, they are formed during flood conditions and are periodically and quite naturally inundated by water.

● During a flood, water containing large quantities of alluvium (river silt) pours out over the flat valley floor.

● The water slowly soaks away, leaving behind the deposited sediment.

● Over hundreds of years, repeated flooding forms a thick alluvial deposit which is fertile and often used for farming.

Floodplains become wider due to the lateral erosion of meanders.

● When the outside bend of a meander meets the edge of the river valley, erosion will cut into it, thereby widening the valley at this point.

● As meanders slowly migrate downstream, the entire length of the valley will eventually be widened.

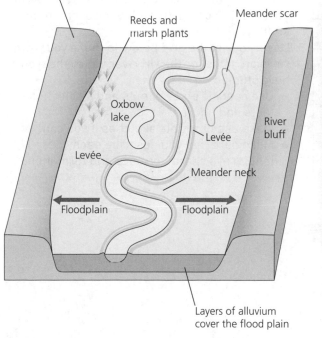

River bluff (area of slightly higher land along the edge of floodplain)
Reeds and marsh plants
Meander scar
Oxbow lake
Levée
River bluff
Levée
Meander neck
Floodplain
Floodplain
Layers of alluvium cover the flood plain

Figure 11.10 Characteristic features of a floodplain

Revision activity

Use a series of simple diagrams to explain the formation of a levée.

Estuaries

In the UK, wide river **estuaries** are commonly formed at river mouths. Estuaries are characterised by extensive deposits of mud, forming mudflats and saltmarshes.

- Incoming tides force seawater and sediment up the river channel, mixing with river water (also carrying huge quantities of sediment) flowing in the opposite direction towards the sea.
- Where the incoming seawater meets the outgoing freshwater, velocity falls dramatically, resulting in significant deposition.
- Over time, these muddy sediments break the water surface to form extensive mudflats.
- Vegetation colonises the mudflats to form saltmarshes.

In some locations, these extensive muddy deposits have been reclaimed and used for large-scale industry or farmland. For example, the River Tees estuary has extensive chemical works and oil refineries constructed on mudflats.

Revision activity

1 Draw an annotated diagram to show how extensive mudflats are formed at river estuaries

2 In the exam you may need to refer to landforms of erosion and deposition in a UK river valley that you have studied. There are many suitable UK river valleys, for example the River Tees in northeast England or the River Severn in the southwest.

3 Use a copy of the grid below to make your own revision notes for the UK river valley that you have studied.
  - Use specific place names and a selection of facts and figures.
  - Make use of simple sketches if you can.
  - Make sure that you describe each landform's characteristics and formation.

My example of a UK river valley:

_____

Landforms of erosion	Landforms of deposition

Exam tip

Focus on the **characteristics** only. You do not need to describe the **formation** of a waterfall and will get no marks for doing so.

Exam practice

1 What are interlocking spurs? (2 marks)
2 Describe the characteristics of a waterfall. (4 marks)
3 Explain how the processes of erosion and deposition are responsible for forming the characteristic features of a meander. (6 marks)
4 With reference to a UK river valley that you have studied, describe the landforms resulting from deposition. (6 marks)

ONLINE

11.3 River management strategies

What are the factors affecting flood risk?

A **flood** occurs when water can no longer be contained within a river channel. The **flood risk** can be increased by physical and human factors.

Physical factors

- **Precipitation** – heavy or prolonged rainfall can saturate soils and make flooding more likely as water flows quickly over the surface.
- Geology – impermeable rocks (rocks that do not allow water to pass through them) result in rapid overland flow, making flooding more likely.
- Steep slopes – water travels fast down steep slopes in mountainous areas, increasing the flood risk.

Human factors

Some land uses increase the flood risk:

- Farming – bare soils can transfer water quickly, particularly when they become saturated in the winter. Ploughing down slopes can create runnels for water to flow down rapidly towards river channels.
- Urbanisation – water can be transferred rapidly over impermeable concrete and tarmac surfaces and through drains and sewers to nearby rivers. The lack of vegetation (grass and trees) results in little rainwater absorption and evaporation.
- Deforestation – this removes the 'umbrella effect' that slows down water transfer and results in much of the precipitation being temporarily stored or used up by trees. As a result, more water reaches river channels, increasing the flood risk.

> **Revision activity**
>
> Draw a spider diagram to show the physical and human factors affecting flood risk.

> **Now test yourself**
>
> 1 Define 'flood risk'.
> 2 Explain how precipitation, geology and land use can affect the flood risk.
>
> TESTED

What are hydrographs?

In describing the response of river **discharge** to a rainfall event, a **hydrograph** helps to understand flood risk. Engineers use hydrograph models to predict likely flooding, enabling warnings to be issued and people to be evacuated. Figure 11.11 shows the main characteristics of a hydrograph.

There are several factors that can affect the shape of a hydrograph.

Type of hydrograph	Physical/human causes
'Flashy' rapid response hydrograph posing a high flood risk	<ul><li>Steep slopes resulting in rapid runoff.</li><li>Impermeable rocks encouraging rapid overland flow.</li><li>Heavy or prolonged rainfall.</li><li>Saturated or frozen soils.</li><li>Deforestation encourages rapid transfer of water to rivers.</li><li>Urbanisation – impermeable surfaces encourage rapid overland flow.</li></ul>
'Flat' slow response hydrograph posing a low flood risk	<ul><li>Gentle slopes slow down water transfer.</li><li>Permeable rocks, allowing water to soak into rocks where transfer is very slow.</li><li>Drizzle</li><li>Deep, dry soils able to absorb water.</li><li>Afforestation resulting in water being intercepted and evaporated.</li></ul>

> **Now test yourself**
>
> 1 What is a hydrograph?
> 2 Explain how precipitation type, intensity and duration can affect the shape of a hydrograph.
>
> TESTED

What are river management strategies?

At the heart of all management is the principle of cost–benefit analysis. Management schemes have economic, social and environmental costs which have to be compared to the benefits offered by protection. Usually it is the economic costs and benefits that dictate whether a management scheme proceeds.

Two types of river engineering can be adopted to prevent flooding: **hard engineering** and **soft engineering**.

Hard engineering

Strategy	Costs	Benefits
Dams and reservoirs – concrete or earth dams can be used to control river flow by creating an artificial lake called a reservoir. Most strategies involve multi-purpose schemes, preventing flooding, providing drinking water and creating an amenity for tourism and recreation.	• High economic costs (e.g. Kielder Dam cost £167 million). • Loss of farmland and buildings flooded by the reservoir. • Social costs include the displacement of people from the flooded area (e.g. Kielder Reservoir led to over 50 families being relocated). • Reservoir silts up over time, limiting its storage capacity. • The dam can affect migrating fish and reduce the beneficial effects of flooding downstream (deposition of fertile silt). • Habitats damaged or destroyed.	• Can provide hydro-electric power, benefiting industry and meeting domestic needs (e.g. Kielder provides electricity for 10,000 people). • Creates new wetland habitats and benefits aquatic ecosystems and fisheries. • Helps control river discharge, preventing flooding downstream. • Creates a source of drinking water. • Provides opportunities for tourism and recreation (sailing, fishing, walking).
Channel straightening – cutting of meanders to create a more efficient channel better suited to the rapid transfer of water.	• Straightening speeds water flow along a river, increasing the flood risk downstream through meandering sections. • This strategy can be expensive and high-maintenance, especially dredging. • Aquatic habitats affected by changes in the velocity characteristics of the channel. • Straightened sections can look unattractive (especially if concrete-lined) and unnatural.	• Effective in speeding up water flow along a short stretch of river, reducing flood risk. • Navigation can also be improved (e.g. the Portrack Cut on the River Tees near Stockport). • Insurance premiums may be reduced for homeowners.
Embankments – river banks can be raised artificially using earth or concrete. This creates a larger cross-sectional area, enabling the river to hold more water.	• Can be expensive, especially when constructed in towns. • Look artificial and unattractive. • Can lead to more serious flooding if the embankment fails.	• The river channel has an increased capacity for carrying water, reducing the flood risk. • Embankments create walkways for local people, they can be grassed and trees planted (e.g. at Bridge of Allen, Stirling). • New river-bank habitats can be created.
Flood relief channels – new channels constructed to by-pass towns or other high-value land.	• Can be very expensive if constructed across high-value land (e.g. Jubilee River at Maidenhead cost £110 million). • Regular maintenance is required to retain channel efficiency – this can be expensive. • Habitats may be disturbed. • If concrete is used, they can look unnatural and unattractive.	• Can be effective in reducing the flood risk in high-value areas, such as town centres. • The new channel provides opportunities for recreation, e.g. walking and fishing. • New aquatic habitats created. • Insurance premiums may be reduced for local people (e.g. Jubilee River has reduced the flood risk for over 3,000 properties).

Soft engineering

Strategy	Costs	Benefits
Flood warnings and preparation – rivers are constantly monitored and computer programs enable scientists to predict possible flooding, issuing warnings and evacuating people.	● Need for monitoring equipment, scientific expertise and computer modelling. ● People may not always respond appropriately, particularly if warnings turn out to be false alarms.	● Focus on behavioural responses and helping people to be prepared and act accordingly. ● Sustainable, recognising that flooding is a natural event and people need to live with floods. ● Ensures safety without high costs of hard engineering.
Floodplain zoning – local planning to restrict land uses in high-risk areas, so that high-value land uses are far away from flood-prone areas.	● Restricts economic development in an area as certain land uses are prohibited. ● Hard to implement retrospectively where urban developments have already taken place on floodplains. ● Current housing shortages will continue if land cannot be used for building. ● Habitats may be destroyed by building elsewhere.	● Low-cost option, based on planning regulations rather than construction. ● Reduces additional impermeable surface coverage of the floodplain. ● Protects and conserves water meadows for wildlife and for people to use for recreation. ● Reduces insurance costs when property is flooded.
Planting trees – afforestation increases the interception of precipitation, reducing the amount of water that might flow into a river. Trees also use up large quantities of water and encourage it to percolate deep into soils and rocks.	● Can reduce habitat (and species) diversity when hillsides are clothed in trees. ● Can lead to increased acidity in soils. ● Loss of potential farmland.	● Benefits wildlife by creating habitats. ● Natural method of slowing down water transfer in the drainage basin. ● Helps to absorb and store carbon, reducing the amount of carbon dioxide in the atmosphere (greenhouse gas). ● Relatively inexpensive.
River restoration – restoring straightened channels to their previous meandering pattern slows down water flow. While small floods may occur in the restored area, the flood risk is reduced downstream.	● Can be expensive to construct new channels and maintain them. ● Change in land use. ● Acceptance that some flooding will occur in the area, which may create inconvenience for local people using the floodplain.	● Recreates a natural attractive fluvial environment, providing a greater variety of aquatic habitats. ● Restores wetland areas, increasing biodiversity on the floodplain. ● Effective in reducing flooding downstream.

Now test yourself

TESTED ☐

Describe how channel straightening, embankments and floodplain zoning help to reduce the flood risk.

Revision activity

Draw a spider diagram (or create flashcards) to act as a revision summary describing the hard and soft engineering options for reducing the flood risk.

Example

Flood management scheme in the UK

Jubilee River – flood relief channel

The Jubilee River is a flood relief channel on the River Thames. It was constructed to reduce flood risk in a number of towns that had experienced expensive flooding in the past. This stretch of the River Thames is highly developed, with extensive housing developments and infrastructure.

The scheme was completed in 2002 and cost £110 million. The new river is nearly 12 kilometres in length and up to 50 metres wide. It is maintained at low flow conditions and takes excess water from the River Thames when necessary.

Social issues	Economic issues	Environmental issues
• Protection of wealthy properties in Maidenhead and Eton at the expense of less affluent areas further downstream at Wraysbury, below the confluence of the Jubilee River and the River Thames. • Paddle boaters, promised a fully navigable river, are disrupted by weirs.	• It was the most expensive flood relief scheme in the UK. • Floods in 2003 damaged weirs – the cost of maintenance is quite high. • Additional flood relief measures are required downstream; the scheme has not so much solved the problem as shifted it elsewhere (serious flooding in 2014).	• Extensive flooding in 2014 below the confluence damaged habitats. • Concrete weirs are unattractive. • Algae collects above the weirs, affecting the aquatic habitat.

Banbury – flood storage reservoir

Banbury is a market town in Oxfordshire with a population of around 45,000 people. It has been affected by flooding of the River Cherwell many times in the past. In 1998 serious flooding closed the railway station and caused damage exceeding £12 million. Further flooding occurred in 2007.

An imaginative scheme to create a flood storage reservoir to the north of the town was completed in 2012 at a cost of £18.5 million. The scheme involved the construction of a 3-kilometre-wide and 4.5-metre-high embankment running parallel to the M40 motorway to create a storage basin capable of holding 3 million cubic metres.

The reservoir is designed to hold back excessive rainwater from entering the river by releasing it slowly through specially designed apertures. Excess water builds up, filling the storage reservoir. This will be gradually released into the River Cherwell in the subsequent days, preventing a surge of water that would otherwise increase the flood risk in the town.

Social issues	Economic issues	Environmental issues
• Transport disruption for local people has been alleviated by the raising of the A361 route into Banbury. • New footpaths and green areas have improved people's lives. • Local people are less concerned about the prospect of future flooding.	• Despite the high cost, benefits (protecting property and transport) are estimated at over £100 million. • Reduced costs for flood damage for homeowners and transport (roads, railway station). • Potentially lower insurance premiums.	• Some habitat destruction in the construction of the embankment. • Biodiversity Action Plan has resulted in planting trees and hedgerows and constructing ponds. • Reservoir will provide a temporary habitat for waterbirds. • The concrete apertures are unnatural in the landscape.

Exam practice

1 Explain how physical factors can increase the flood risk. (6 marks)
2 What is the discharge of a river? (2 marks)
3 Explain the benefits of soft engineering strategies in reducing the flood risk. (6 marks)
4 Using an example and with reference to economic and environmental issues, justify the implementation of a flood management scheme. (6 marks)

ONLINE

Exam tip

With Question 4, you should refer to an example and focus on economic and environmental issues. You must attempt to justify the flood management scheme on these grounds, so make clear connections between the issues and the scheme.

12 Glacial landscapes in the UK

12.1 Glacial processes

How did the last ice age shape the UK?

REVISED

Glacial processes have been responsible for sculpting some of the UK's most dramatic landscapes. Figure 12.1 shows the maximum extent of ice across the UK at the height of the last ice age, some 25,000 years ago.

- Vast ice sheets spread over the UK from the north and northeast to cover all of northern Wales and northern England (previously, ice had advanced further south but southern England was never covered by ice).
- Huge tongues of ice called glaciers flowed out of mountainous areas.
- Unglaciated areas in the south experienced frozen conditions (permafrost).

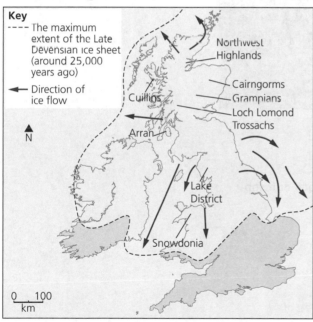

Key
- - - - The maximum extent of the Late Devensian ice sheet (around 25,000 years ago)
⟵ Direction of ice flow

N

Northwest Highlands
Cuillins
Cairngorms
Grampians
Loch Lomond Trossachs
Arran
Lake District
Snowdonia

0   100
km

Figure 12.1 Glaciated upland areas in the British Isles (UK and Eire)

What are glacial processes?

REVISED

Freeze-thaw weathering

Freeze-thaw weathering is an active process of weathering in glacial environments where rocks are exposed to changing temperatures. It would have been very active in periglacial (edge of ice) conditions, before and after ice covered an area, as well as on mountain peaks exposed above the ice.

As seen in Figure 12.2, freeze-thaw weathering requires:
- frequent temperature changes above and below freezing (0°C) – to enable freezing and thawing to occur
- presence of liquid water
- presence of rocks with cracks/holes

The angular rock fragments become powerful tools of erosion when trapped beneath glaciers (see abrasion, page 78).

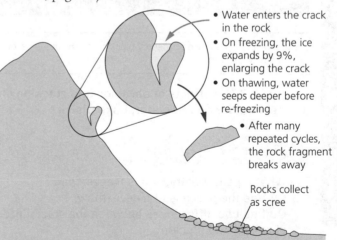

- Water enters the crack in the rock
- On freezing, the ice expands by 9%, enlarging the crack
- On thawing, water seeps deeper before re-freezing
- After many repeated cycles, the rock fragment breaks away

Rocks collect as scree

Figure 12.2 Freeze-thaw weathering

> **Revision activity**
>
> Draw a sequence of labelled diagrams from memory to describe the process of freeze-thaw weathering.

Glacial erosion

Abrasion and **plucking** are two important processes of glacial erosion (Figure 12.3).

- Abrasion – angular rocks trapped beneath the glacier which scratch and smooth the underlying bedrock. The presence of meltwater beneath the glacier is important because it helps to lubricate the ice, enabling it to grind its way forward.
- Plucking – meltwater beneath the ice freezes and bonds pieces of loose bedrock to the glacier. When the glacier moves forward, these loose pieces of rock are plucked away from the bedrock, leaving behind a very jagged and angular surface.

Smooth abraded rock with scratches (striations) in the direction of the ice flow

Jagged, angular surface formed by plucking on the down-valley side of the rock outcrop

Figure 12.3 The effects of abrasion and plucking on a rock outcrop (roche moutonnée), Honister, Lake District

Now test yourself

TESTED

1 What type is weathering is freeze-thaw?
2 Outline the process of freeze-thaw.
3 Outline two differences between the glacial processes of abrasion and plucking.

Glacial movement and transportation

Unlike rivers or the sea, it's impossible to actually see ice moving, yet it does so incredibly slowly. In response to the huge mass of ice – glaciers may be well over 1,000 metres thick – and gravity, the ice within a glacier does move forward in the following ways:

● Basal slip – meltwater beneath a glacier enables it to slide forward as a mass by a few metres a year.
● Internal deformation – this involves the slipping and deformation of individual ice crystals within the glacier, rather like the movement of individual grains of sugar in a pile of granulated sugar!

Glaciers can advance or retreat (most are now retreating in response to climate change):

● Glacier advance – this occurs when the amount of additional snow and ice (accumulation) in a year exceeds the amount of melting (ablation).
● Glacier retreat – this occurs when the amount of ablation exceeds accumulation.

When a glacier moves forward it can act like a giant earthmover, **bulldozing** piles of rock debris in front of it to create a high ridge called a moraine. The furthest advance of a glacier is marked by a terminal moraine (see page 83).

Smaller glaciers in mountainside depressions may move by **rotational slip**, a curved movement that results in the formation of a corrie (see page 68).

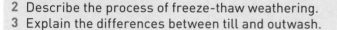

Now test yourself

TESTED ☐

What is the difference between bulldozing and rotational slip?

Glacial deposition

Glaciers act like giant conveyor belts transporting vast amounts of weathered and eroded sediment.

● When melting takes place towards the snout of a glacier, the sediment is dumped on the ground to form a deposit called **till**. Till is characterised by having a range of sizes (it is poorly sorted) and mostly angular rocks, reflecting the lack of transport by water.
● Meltwater pouring out from the snout of a glacier will carry sediment away, depositing it as a vast **outwash** plain in front of the glacier. Carried by water, this sediment is better sorted and the particles are more rounded due to the process of attrition (see page 68).

Exam practice

1 Study Figure 12.1 on page 77. Describe the extent of ice cover across the UK. (4 marks)
2 Describe the process of freeze-thaw weathering. (4 marks)
3 Explain the differences between till and outwash. (4 marks)

 ONLINE ☐

12.2 Glacial landforms

What landforms result from erosion?

Corries

Corries are characteristic landforms of mountainous areas. They are formed by a combination of processes including freeze–thaw weathering, plucking and abrasion (Figure 12.4). Rotational slip causes the mountainside hollow to be scooped out by the ice, which is at its thickest and most powerful at this point. This explains its depth and often accounts for the presence of a corrie lake (tarn) trapped within the bowl of the corrie. A reduction in ice thickness (making it less erosive) explains the presence of a rock lip at the front of the corrie.

> **Revision activity**
>
> Practise drawing the diagram of corrie formation in Figure 12.4 and add detailed labels to describe the formative processes.

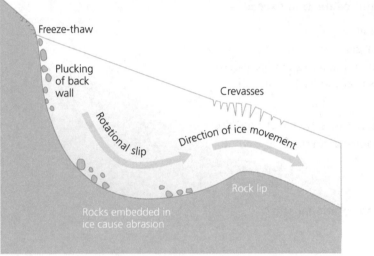

Figure 12.4 The formation of a corrie

Arêtes and pyramidal peaks

Arêtes and **pyramidal peaks** are also characteristic landforms of the high mountains (Figure 12.5).

- Arêtes – when two corries are eroded back to back, the ridge in between them becomes narrower until it may be just a few metres across.
- Pyramidal peak – when several corries erode back into a mountain, a pronounced peak may be formed, such as the Matterhorn in the European Alps.

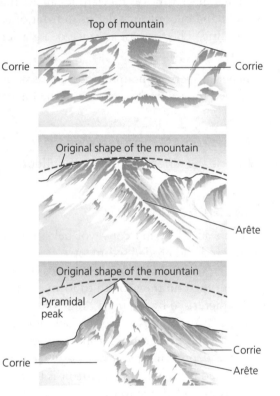

Figure 12.5 Arêtes and pyramidal peak

> **Now test yourself**
>
>
> In no more than twenty words for each, write a definition of a corrie, an arête and a pyramidal peak.

Glacial troughs and associated features

Glaciers are enormously powerful. They can carve huge U-shaped valleys called **glacial troughs**. They do this by being so vast and solid that they cut through interlocking spurs rather than flowing around them (as a river would do). The tips of these spurs are chopped off (through abrasion, see page 78) to form vertical cliffs called **truncated spurs**.

Smaller glaciers in tributary river valleys also form glacial troughs but on a smaller scale. With less downward erosion, when the ice finally melts, these tributary valleys are left at a higher level than the main trunk valley. These smaller valleys are called **hanging valleys**. They are often marked by waterfalls plunging over the valley wall.

Ribbon lakes are long, narrow and often very deep freshwater lakes. They are common glacial landforms found on the floor of glacial troughs – in fact, their abundance accounts for the name Lake District in north-west England!

Ribbon lakes result from a localised increase in vertical erosion, often involving rotational slip scooping out the bedrock. Increased erosion at this point may occur:
- where a band of weaker, more easily eroded rock crosses the valley
- where a tributary glacier joins the valley, increasing the mass of the ice and resulting in greater erosion
- where the valley sides become narrower, increasing the depth and power of the glacier.

Figure 12.6 describes the landforms associated with glacial troughs.

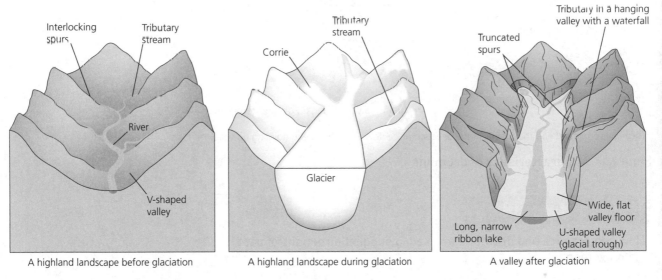

A highland landscape before glaciation   A highland landscape during glaciation   A valley after glaciation

Figure 12.6 Landforms associated with glacial troughs

Now test yourself

TESTED ☐

1 Describe the characteristics of a truncated spur.
2 What is the difference between a glacial trough and a hanging valley?
3 Explain why a ribbon lake may form on the floor of a glacial trough.

What landforms result from transportation and deposition?

Erratics

Erratics are rocks that are in the wrong place! For years, nobody understood why rocks of one type appeared in an area made out of completely different rock. It is now understood that these 'alien' boulders were in fact transported by glaciers and ice sheets that have long since melted.

Once the source location of an erratic has been pinpointed, scientists can use this to suggest the direction of movement of past ice sheets and glaciers.

Drumlins

Drumlins are 'egg-shaped' hills often occurring in groups called swarms (Figure 12.7). A drumlin can extend for several hundred metres and can have a height of 10 metres or more.

Drumlins are made of till (see page 79). It is thought that drumlins are moulded and streamlined by the ice as it moves over a considerable thickness of till. Drumlin formation remains a bit of a mystery because, beneath hundreds of metres of ice, it is impossible to see them being formed! They are, however, useful because they indicate past directions of movements.

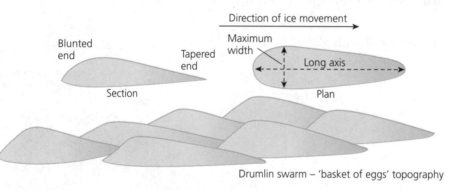

Drumlin swarm – 'basket of eggs' topography

They have a tapered end that points in the direction of ice flow and a blunt end pointing up-valley

Figure 12.7 Characteristics of drumlins

How can erratics and drumlins be used to suggest directions of past ice movement?

Moraines

Moraines are formed by the deposition of poorly sorted, angular till deposits carried by the ice, and then dumped when the ice melts. Figure 12.8 shows the characteristics and location of the four main types of moraine.

● Lateral moraine – an elongated ridge of till that builds up at the edge of a glacier where it meets the valley side. It is continually fed by rocks from above, loosened by freeze-thaw (see page 77).

● Medial moraine – when two tributary glaciers meet, two lateral moraines join together to form a single ridge in the centre of the main glacier.

● Ground moraine – the uneven till deposits smeared on the bedrock beneath a glacier.

● Terminal moraine – often resulting from bulldozing (see page 79), this extensive ridge, at right angles to the valley, forms at the snout of a glacier. It marks its furthest extent. When glaciers retreat, they may remain stationary for some time, or even advance slightly, allowing smaller recessional moraines to build up.

When the ice melts, the moraines are left behind (though they often become eroded by meltwater, so may be hard to identify). Lateral moraines on the edges of valleys often remain largely intact, as they are less likely to be eroded away by meltwater.

Revision activity

Draw your own summary diagram to show the location of the four main types of moraine – add recessional moraine if you like. Add text boxes to describe the formation of each one.

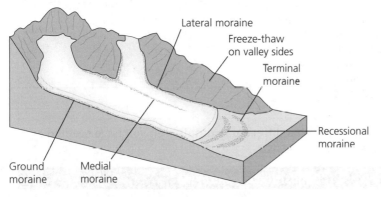

Figure 12.8 Types of glacial moraine

Exam practice

1 What is an arête? (2 marks)
2 Explain how the processes of glacial weathering and erosion are responsible for the formation of a corrie. (6 marks)
3 Describe the characteristics of drumlins. (4 marks)
4 With reference to an upland area in the UK, describe the landforms of glacial erosion **or** deposition. (6 marks)

Note. You will need to read page 84 before attempting Question 4.

ONLINE

Exam tip

When you are asked to describe characteristics of a landform, remember that you only need to write about how they look (not how they are formed).

Exam tip

When a question asks you to choose something to write about ('X **or** Y'), remember to choose one – not both!

Major landforms resulting from erosion and deposition

The table below illustrates the level of detail that you should aim to have for your chosen case study. It will often be the case that there will be more to write about landforms of erosion than landforms of deposition. This reflects the fact that your case study needs to come from an upland area of the UK and also that landforms of erosion tend to be larger and more distinctive than landforms of deposition.

Cadair Idris, Snowdonia, North Wales

Erosional landforms	Depositional landforms
• Corrie – the corrie containing the lake (tarn) Llyn ('llyn' is Welsh for 'lake') Cau is a steep-sided corrie facing towards the east. It is located on the southern side of the Cadair Idris mountain range and occupies an area of about 1 kilometre². • Arêtes – there are several arêtes in the area marking the edges of corries, including the one containing Llyn Cau. They are marked by cliffs so are extremely steep and narrow. One example is Craig Cwn Amarch. • Pyramidal peak – there is a pyramidal peak at the back of the corrie containing Llyn Cau at 791 metres above sea level. • Glacial trough – a broad glacial trough extends from northeast to southwest to the south of Cadair Idris. It is about 500 metres across, flat-bottomed and with very steep sides. It contains a ribbon lake (Tal-y-llyn), which is about 1.5 kilometres in length. There are several examples of hanging valleys and truncated spurs alongside this valley.	Remember that depositional landforms tend to be concentrated in lowland areas where the ice melts and often they are washed away by meltwater streams. • The Tal-y-llyn valley exhibits a number of depositional landforms although they are rather indistinct. Much of the valley floor is covered by till in the form of hummocky ground moraine. There may also be the remnants of medial moraines. On the valley sides are isolated remains of lateral moraines. • Throughout the area, there are boulder erratics transported by the ice from elsewhere and deposited on bare rocky surfaces.

Use the blank table below either to write revision notes on a different example that you have studied or to add more detail (such as simple sketch maps) on Cadair Idris.

Erosional landforms	Depositional landforms

Remember that you could well be asked to write about 'an example of an upland area in the UK' so you must learn a few place names and details of landforms. Remember that you only need to learn **one** example.

12.3 Land use in glaciated upland areas

What economic activities are carried out in glaciated upland areas?

Farming

Opportunities for agriculture in upland glaciated areas are limited:

● Soils tend to be thin and infertile in the mountains, and can be waterlogged in valley bottoms.
● Slopes are steep and often covered by scree or bare rock outcrops.
● The climate is harsh, with high rainfall, low temperatures and strong winds. Heavy snowfalls can occur in the winter.

Despite these challenging conditions, extensive farming involving sheep grazing is widespread with some crops, primarily grass for hay/silage (grass compacted and stored in airtight conditions to be used as animal feed in the winter). Some specialist farming, such as Highland beef cattle and deer (for venison meat), exists in places.

Forestry

Commercial forestry is widespread throughout upland glaciated areas, mostly involving conifer plantations. These fast-growing trees are well suited to the harsh climate and can thrive on the thin, acidic soils. Trees are used to provide wood for furniture and construction. Some are chipped to use as biofuel.

Quarrying

Upland glaciated areas are made of tough rocks that can be quarried and used for a variety of purposes, such as road building:

Lake District slate	This distinctive blue-grey rock is used around the world as a roofing and decorative material. The Lake District has thirteen active quarries.
Pennines limestone	Limestone is a widely used building material. Limestone fragments and gravel are a popular landscaping material for gardens.
Highlands granite	This tough, resistant rock has a range of uses, from pavement edges to kitchen work surfaces. Granite from the glaciated Scottish island of Ailsa Craig is used for the 'stones' in the sport of curling, due to its unusually uniform hardness.

Tourism

Tourism is a hugely important economic activity, providing thousands of jobs and contributing greatly to local economies through shops, hotels, restaurants and visitor attractions. Vast numbers of people are drawn to upland glaciated areas such as the Lake District and Snowdonia for hiking, cycling, climbing and nature watching.

Aviemore (Cairngorms) is a multi-activity centre offering opportunities for mountain biking, hiking and skiing in the winter. A wildlife park, steam railway and shops draw many thousands of people to this mountain resort each year.

> **Revision activity**
>
> Create a spider diagram to summarise the opportunities for economic activities in an upland glaciated area.

Figure 12.9 Tourism in Aviemore

What are land use conflicts in glaciated upland areas?

Land use conflicts can arise between different land users in upland glaciated areas. Here are some examples:

- Farming and tourism – tourists use footpaths that cross farmland, occasionally leaving access gates open through which farm animals can escape. Loose dogs can worry sheep. Some people leave litter (e.g. plastic, drinks cans and glass), which can be dangerous to animals.
- Quarrying and **conservation** – quarrying can damage the natural environment, destroy habitats and make the landscape unattractive. Lorries on narrow country roads can cause damage to verges and create air pollution.
- Tourism and conservation – intensive tourism can lead to pollution (air pollution from vehicles, litter) and footpath erosion.

Conflicts can exist between economic development and conservation:

- Energy developments – glaciated upland areas offer some ideal locations for developing wind farms. Land prices are relatively low, population is sparse and exposed locations receive strong winds. However, local people are concerned about the impact of wind farms on the natural environment (visual pollution). Local businesses relying on tourism are concerned about the possible negative impact on tourism as landscapes become scarred by clusters of wind turbines.
- Reservoir construction – upland glaciated areas can be ideal locations for reservoirs, with high rainfall totals, deep valleys and a low population density. However, reservoirs can destroy habitats, impact river regimes and disrupt water flow. They can, however, bring economic benefits and create tourist amenities (fishing, sailing). They can also reduce the flood risk.
- Forestry – conifer plantations can have a significant impact on the natural environment. Few species of tree are planted, leading to a reduction in biodiversity. As they grow, they block out light, restricting the range of flowering plants on the forest floor. With limited vegetation, plantations are often devoid of wildlife.

> **Revision activity**
>
> Use your textbook to write short revision notes on one example of a conflict between development and conservation (e.g. the Glenridding zip wire in the Lake District or hunting in Scotland).

> **Now test yourself**
>
> 1 Define the term 'conservation'.
> 2 Describe one example of a conflict between development and conservation.
>
> TESTED

Example

Tourism in an upland glaciated area: the Lake District

You only need to be able to refer to **one** upland area, which could be the
Lake District **or** the Isle of Arran.

Attractions for tourists

- Opportunities for boating, fishing and hiking along the shores of ribbon lakes such as Ullswater and Windermere.
- Mountain landscapes (e.g. Helvellyn) offer opportunities for hiking and climbing.

- Gorge scrambling, abseiling and off-road driving are some adventure activities offered.
- Small towns (such as Grasmere and Ambleside) and historic houses and gardens (e.g. Beatrix Potter's house) are popular attractions.

Impacts of tourism

Impact	
Social	• Almost 50 million people visit the Lake District each year, putting a huge amount of pressure on roads and other facilities. • Congestion is a major issue on the narrow and winding roads. • Local people are unable to afford to buy houses due to purchase of second homes by wealthy tourists. • Jobs tend to be poorly paid and seasonal.
Economic	• Employment in tourism for thousands of people boosts family incomes and provides money to spend in the local economy. • Businesses, such as adventure tourism, are promoted by tourism. • Tourists spend over £1,000 million each year in hotels, restaurants and shops, boosting local economies.
Environmental	• Pollution (traffic, litter, etc.) can be an issue. • Walkers can create conflicts with farmers if gates are left open or dogs worry livestock. • Footpath erosion is a major issue in some popular locations.

Managing the impact of tourism

Impact of tourism	Management strategies
Traffic congestion	• Public transport is encouraged, with special bus routes serving hikers and cyclists, e.g. Honister Rambler. • Traffic calming in villages. • Some settlements (e.g. Ambleside) identified as 'transport hubs', with co-ordinated facilities, such as car parks, bus stops, footpaths and cycleways.
Footpath erosion	• Volunteer groups and organisations such as 'Fix the Fells' work with local landowners to restore and repair footpaths, often using local stone and even sheep's wool to make them more resilient (Figure 12.10). • Improved signage and encouraging people not to stray off dedicated footpaths. • Re-planting of native plants and those able to withstand trampling reduces erosion.

Figure 12.10 Tackling footpath erosion in the Lake District

Tourism in an upland glaciated area: the Isle of Arran

You only need to be able to refer to **one** upland area, which could be the Lake District **or** the Isle of Arran.

Attractions for tourists

- Spectacular glacial scenery attracts walkers, climbers and cyclists.
- Goatfell (874 metres) is the Isle of Arran's highest mountain and one of the most popular visitor attractions.
- Climbers enjoy the challenges of A'Chir ridge, a knife-edged arête separating two corries.
- Opportunities exist for adventure activities such as abseiling and paragliding as well as helicopter tours.

Impacts of tourism

Impact	
Social	• Congestion is a major issue on the narrow and winding roads, causing conflicts between tourists and local people. • Many jobs have been created but they tend to be seasonal • Some younger people are moving to the island, creating a more balanced population structure.
Economic	• Over 200,000 people travel to the Isle of Arran each year, using hotels and restaurants and contributing to the local economy. • Tourism generates an estimated £30 million annually. • New whisky visitor centre has opened at Lochranza.
Environmental	• Footpath erosion is a major issue (e.g. Goatfell), made worse by the heavy rainfall. • Pollution (traffic, litter, etc.) can be an issue. • Walkers can create conflicts with farmers if gates are left open or dogs worry livestock.

Managing the impact of tourism

Impact of tourism	Management strategies
Seasonality of employment	• New all-season attractions have opened, such as the Balmichael Centre and Auchrannie Resort, to encourage tourism in the winter. • New website 'Visit Arran' encourages people to visit out of season for mini-breaks.
Footpath erosion	• National Trust for Scotland organises restoration work, stabilising paths using local materials and creating steps to improve access and safety. • Native plants are re-planted to help stabilise soils and reduce the impact of heavy rainfall.
Accidents	• Arran Mountain Rescue Team provides support and emergency rescue.

Now test yourself

TESTED

For an example of an upland glaciated area, outline **one** social, **one** economic and **one** environmental impact resulting from tourism.

Exam practice

1 How do glaciated upland areas provide opportunities for economic activities? (4 marks)
2 Explain why conflicts might exist between development and conservation in upland glaciated areas. (6 marks)
3 Use an example of an upland glaciated area to describe strategies used to manage tourism. (6 marks)

ONLINE

13 Urban growth across the world

13.1 Global pattern of urban change

The world is becoming increasingly urbanised with more and more people living in towns and cities. In 1950, 30 per cent (746 million people) of the world's population was urban. By 2014 this figure had risen to 54 per cent (3.9 billion people) and it is projected to rise to 66 per cent (6.4 billion people) by 2050. **Urbanisation** is one of the most important and challenging trends for the future, creating many issues in both cities and the countryside.

What are urban trends in different parts of the world?

Figure 13.1 is a choropleth map showing the global pattern of urban population. Notice that there are huge variations between richer HICs and poorer LICs.

- Richer HICs (e.g. the USA, much of Europe, Japan and Australia) – these countries are highly urbanised, with 75 per cent of the population living in urban areas. North America is the most urbanised region, with 82 per cent of the population living in urban areas. The figure for Europe is 73 per cent (the UK is 82 per cent). Urbanisation in these regions is slowing down. In some countries there is a reverse flow of people (counter-urbanisation) who have decided to leave crowded cities to live in the countryside.

- Poorer LICs (e.g. much of Africa and some parts of Asia and the Middle East) – these are still predominantly rural, with Africa 40 per cent urban and Asia 48 per cent rural. However, it is these regions that are experiencing the fastest rates of urbanisation, with Africa expected to reach 56 per cent and Asia 64 per cent by 2050. It is in these regions where the greatest challenges lie for the future.

Just three countries – India, China and Nigeria – are expected to account for 37 per cent of the growth in the world's urban population between 2014 and 2050.

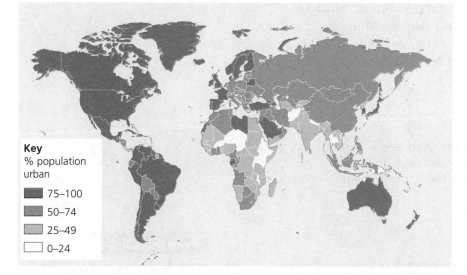

Key
% population urban

- 75–100
- 50–74
- 25–49
- 0–24

Figure 13.1 Global urban population, 2014

What factors affect the rate of urbanisation?

Urbanisation is the result of migration, as people move from the countryside into the cities, and **natural increase**, as birth rates exceed death rates in cities.

Migration

Migration from the countryside to the cities usually results from people considering 'push' and 'pull' factors:

Push factors (disadvantages of living in the countryside)	Pull factors (attractions of living in cities)
• Poor harvests, resulting in shortages of food. • Limited opportunities for well-paid employment. • Limited services (water, electricity, schools, health care). • Issues associated with climate change, such as desertification, soil erosion, floods and drought. • Poor transport infrastructure.	• Prospect of better paid employment. • Large market for goods and services, e.g. street vendors selling food or providing services. • Better schools and health care provision. • Better public transport facilities and access to services such as water and electricity. • Friends and family already living in urban areas encourage others to join them.

Natural increase

Many people living in towns and cities, particularly new migrants, are aged 18–35. This leads to a high birth rate and a high rate of natural increase. As more young people move into urban areas, this trend is likely to increase as the population structure becomes ever more youthful.

What are megacities?

One of the recent trends associated with urbanisation has been the growth of **megacities**. These enormous, sprawling urban areas have grown rapidly in size and number in recent years:

- In 2015 there were 28 megacities. There are expected to be over 40 by 2030.
- Currently Tokyo is the largest city in the world, with a staggering 38 million people, followed by Delhi (25 million), Shanghai (23 million) and Mexico City, São Paulo and Mumbai (21 million).
- While the growth of megacities in HICs (e.g. Tokyo and Los Angeles) is slowing down, those in parts of Asia and Africa (e.g. Mumbai, Lagos and Jakarta) are growing rapidly.

Now test yourself

TESTED

1 Define the following terms as concisely and accurately as you can: urbanisation; migration; natural increase; megacity.
2 What are the recent trends with the size and number of megacities?

Exam practice

1 Study Figure 13.1. Describe the pattern of urban population. (4 marks)
2 Define the term urbanisation. (2 marks)
3 Explain the causes of urbanisation. (6 marks)

ONLINE

Revision activity

Create a colourful revision diagram to show push and pull factors. Can you add any additional factors to your diagram?

13.2 Case study: urban growth in Lagos

This book uses Lagos in Nigeria and Rio de Janeiro in Brazil as case studies of major cities in an LIC or NEE. You only need to study **one** case study – one of these or another one that you have studied at school. If you studied a different LIC or NEE city, you might like to use these chapters (13.2 and 13.3) to guide you in making your own set of revision notes.

What is Lagos like?

With a population of at least 15 million people, Lagos is the most populous city in Nigeria. It is located in the southwest of Nigeria on the coast of the Gulf of Guinea (Figure 13.2).

Under British colonial rule, Lagos was the capital of Nigeria. It became a major trading port and remained the capital city after independence in 1960. In 1991, the Nigerian government established Abuja as the new capital city to encourage development in the interior of the country. Despite this, Lagos remains Nigeria's most important city.

Regional importance	National importance	International importance
• Lagos is important in its provision of schools, universities and hospitals as well as many opportunities for employment, leisure and recreation. • There is a thriving arts and cultural scene in Lagos. • The city is a transport hub, with an international airport and important docks providing raw materials for local industries and exporting products.	• About 80% of Nigeria's industry is located in Lagos and the city generates about 25% of Nigeria's gross domestic product (GDP). • 80% of Nigeria's imports and 70% of its exports pass through the docks. • Lagos is Nigeria's media centre, with many newspapers and television channels operating in the city. It is the centre of Nigeria's film industry. • Home to most banks, financial institutions and the stock exchange, Lagos is Nigeria's finance centre. • Most large companies and Transnational Corporations (TNCs) have their headquarters in Lagos.	• Lagos has one of the highest standards of living in Nigeria and Africa. • Lagos' Apapa port is the fifth busiest in West Africa. • Lagos has been the venue for major sporting events, such as the African Cup of Nations tournament. • The Lagos International Trade Fair, held in November each year, has become a major international business forum, attracting representatives from across the continent. • Lagos is the ICT centre of West Africa, with the largest market on the continent.

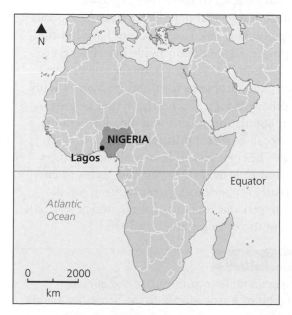

Figure 13.2 Location of Lagos

Describe the location of Lagos.

Create a spider diagram to summarise the regional, national and international importance of Lagos.

How has Lagos grown?

Lagos has grown rapidly to reach its current population of about 15 million (see Figure 13.3), sprawling into the surrounding countryside. Initially a fishing village, Lagos developed into a thriving colonial sea port as international trade developed. Since the 1970s, its oil boom has drawn many thousands of people to the city.

Essentially, population growth is due to the following:

- Migration – rural–urban migration has been the main cause of Lagos' growth during the last 50 years. This has been driven by push factors (poor rural services, low wages, land shortage and climate change) and pull factors (higher paid job opportunities, and better education and health care).
- Natural increase – the high rate of migration has resulted in a youthful population in Lagos, which in turn has resulted in a high rate of natural increase, due to the relatively high birth rate.

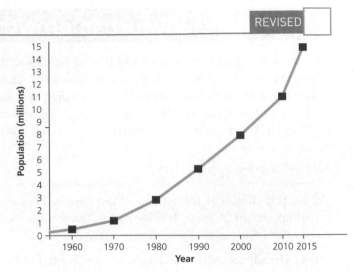

Figure 13.3 Population growth in Lagos

Now test yourself

Explain how migration and natural increase have led to the rapid growth of Lagos' population.

What opportunities are associated with the growth of Lagos?

The growth and development of Lagos as a modern city in Nigeria has created a number of **social opportunities** and **economic opportunities** to improve people's quality of life.

Access to services (health and education)

- Health care – throughout Nigeria, health care is generally underfunded, understaffed and underequipped. In Lagos, health care is better than in the countryside, with greater access to doctors, hospitals and clinics. Health care is divided into public and private sectors – though private hospitals are not necessarily better!
- Education – state schools are operated by the Lagos State Government, which offers all children a basic education focusing on the first nine years. There are several federally funded schools, most of which are boarding schools. Lagos is home to many universities and training colleges (e.g. Yaba College of Technology, which caters for 16,000 students, the Federal College of Fisheries and Marine Technology, and Lagos University Teaching Hospital).

Access to resources (water supply and energy)

- Water supply – Lagos offers a reasonably reliable water supply. Most people dig wells or boreholes to access water underground. Others buy water from street vendors. The Lagos Water Corporation claims to supply over 12 million people with drinking water. The Water Master Plan (2020) aims to meet rising demand by constructing several new water treatment plants.
- Energy – many rural areas do not have lighting and power. Lagos does provide better access to these, although there are frequent power cuts, which cause problems for industry, water supply and other services. Currently, 80 per cent of the urban population relies upon diesel generators, a main cause of air **pollution**. 'Future Proofing Lagos – Energy Sector' is an ambitious project that aims to provide 100 per cent access to energy (currently 60 per cent), street lighting for all commercial and residential areas, and at least 20 per cent use of renewables by 2030.

Revision activity

Construct a table to summarise the social opportunities associated with the growth of Lagos.

Improved economic development

Most industrial areas in Lagos are on the mainland, with good access to the port, or alongside transport arteries, such as railways and main roads. The thriving industrial sector provides opportunities for employment for newcomers.

The manufacturing industrial sector in Lagos is dominated by food and beverages, pharmaceuticals and vehicles. As the wealth of the city increases, so the market is expected to grow, and this will generate further economic development.

The growth of commercial and industrial zones in the Ikeja district in Lagos has led to considerable economic development. Industries have been attracted into the area and some houses have been turned into offices. The district is home to the main international airport and most roads are in good condition. The district also boasts several international hotels, new shopping malls and extensive opportunities for entertainment.

The Lagos State Government expects that future improvements in transport infrastructure and electricity supply will create major economic development.

Now test yourself

How can industrial areas be a stimulus for economic development? TESTED

What challenges are associated with the growth of Lagos?

REVISED

The population of Lagos is expected to double by 2050. The physical growth of the city, together with the poverty of most people there, will create some enormous challenges.

Managing urban growth (slums and squatter settlements)

The lack of affordable property in Lagos has forced millions of people to build their own homes – often no more than shacks – on land they do not own. These vast and sprawling **squatter settlements** tend to be located on marginal land (e.g. marshland) that nobody else wants to build on. Squatter settlements, also called slums, shanty towns and informal settlements, are found in cities throughout the developing world.

- Most homes are constructed from waste materials such as corrugated iron, wooden planks and even cardboard.
- 75 per cent of households occupy a single room.
- Over 50 per cent of households lack a kitchen, bath or toilet.
- Only 11 per cent have access to safe piped water, with the majority using wells or boreholes where the water is often contaminated.

Makoko is a squatter settlement on the edge of Lagos Lagoon. The shortage of land means that some houses are constructed on stilts in shallow water. Most people in Makoko work in the informal sector (paying no taxes) or are involved in fishing.

In 2012 the local authorities started to demolish parts of the squatter settlement. They want to turn the area into a grander 'Venice of Africa'. Local people – despite the difficulties – are determined to remain there, as they have nowhere else to live.

Now test yourself

Suggest the main challenges facing the management of squatter settlements.

TESTED

The provision of clean water, **sanitation** and energy is a major issue in Lagos.

Providing clean water, sanitation and energy

Water supply
- Only 10 per cent of the population has access to safe piped water supplies. The vast majority dig wells or boreholes to extract water from groundwater aquifers, and some buy water from street vendors. This water is untreated and may be contaminated with pollutants.
- In 2012, the Lagos State Water Regulatory Commission began the immense job of trying to regulate street vendors, and license boreholes to try to provide all people with safe water.

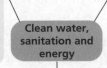

Clean water, sanitation and energy

Sanitation
- Most people have no access to flush toilets. Instead, they use pit latrines, where sewage either drains into the soil (potentially polluting groundwater) or pours into open drains and rivers.
- Lagos Lagoon and its many creeks are heavily polluted with raw sewage.
- Providing people with proper sanitation is an enormous issue, particularly in the densely packed squatter settlements.

Energy
- Despite Nigeria's vast reserves of oil, energy is a big issue in Lagos (see page 92). Most large organisations rely upon back-up generators to keep the lights on.
- New power stations are planned, including one powered by methane from the Olususun landfill site.

Figure 13.4 Clean water, sanitation and energy issues in Lagos

Providing access to services (health and education)

Health care
- This is available, but is not always free, and there may be long queues to see a doctor.
- Vaccinations are available for children but they usually need to be paid for.
- Currently, investment in health care is not keeping up with population growth. Residents are at risk from contagious diseases such as malaria and typhoid.
- Many wealthier people seek medical help abroad.

Access to services

Education
- The government offers free education for all younger children; however, many children in the poorest areas have to work to provide money for their families.
- Secondary schools are limited and are mostly private, and there are not enough universities for the city.
- Industries are being encouraged to work with universities to create relevant courses to create employment opportunities.
- The use of e-learning in the future could extend education to more people.

Figure 13.5 Access to health and education services issues in Lagos

Reducing unemployment and crime

Unemployment
- Unemployment is relatively low, just below 10 per cent.
- However, with no unemployment benefit, people are forced to earn money either legally by paying taxes in the formal sector or illegally within the so-called informal sector. These informal jobs (employing about 40 per cent of the workforce) include street vending, car washing and **waste recycling** (e.g. the Olususun landfill site). They are poorly paid, unregulated and often dangerous.
- In 2016 the Employment Trust Fund provided loans to help people become self-employed.

Unemployment and crime

Crime
- Crime rates in Lagos are high, particularly those involving drugs, vandalism and theft. Rates of assault and armed robbery are high, as are bribery and corruption.
- Kidnapping is a threat, particularly for foreigners. Cyber-fraud and scams are becoming an increasing problem within Lagos' financial community.
- Violent clashes occasionally break out between street gangs known as 'Area Boys'.

Figure 13.6 Unemployment and crime issues in Lagos

Now test yourself

TESTED

1 Outline the management issues associated with providing water and education in Lagos.
2 What are the issues associated with reducing crime?

Now test yourself and Exam practice answers at **www.hoddereducation.co.uk/myrevisionnotes**

Managing environmental issues

- Waste disposal – the city authorities collect just 40 per cent of the 10,000 tonnes of waste produced daily. The waste is dumped in huge landfill sites such as Olusosun, which is in the heart of the city. Only 13 per cent of the waste is recycled, mostly by the 500 people who work informally at the site collecting and then selling materials, such as plastic bottles and clothing. Waste piles up in many areas of the city, particularly in the poorer areas.
- Pollution – air pollution is five times higher than the international recommended limit, mainly because of the poorly maintained and unregulated vehicles that choke the city's roads. Water pollution is a major problem due to domestic sewage and industrial waste (see page 94).
- **Traffic congestion** – Lagos is one of the most congested cities in the world, with the average commuter spending three hours in traffic every day. Congestion results in high levels of air pollution and causes businesses to lose money. The Lagos Metropolitan Area Transport Authority (2003) has introduced a bus rapid transit system (Figure 13.7) with a separate bus lane that transports 200,000 commuters each day to the central business district (CBD) on Lagos Island. A large fleet of minibus taxis ('danfos') operate in the city, but these are often very overcrowded. In 2016, a new light railway opened (Figure 13.7), and further rail routes are planned as part of the Strategic Transport Master Plan.

This plan aims to create:
- ○ an **integrated transport system** linking road, rail and waterway
- ○ a new waterway network using ferries to transport commuters
- ○ more efficient roads with dedicated bus lanes and fewer street markets
- ○ mixed-use developments, with commercial and residential uses so that people have less far to commute
- ○ a new airport on Lekki peninsula
- ○ improved opportunities for walking and cycling.

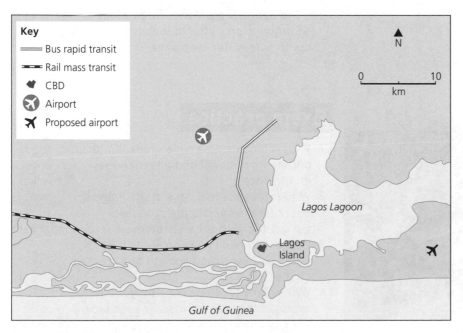

Figure 13.7 New and planned transport developments in Lagos

Now test yourself

Outline the management strategies for reducing transport congestion in Lagos.

TESTED ▢

How urban planning helps the urban poor

The urban poor living in Lagos' squatter settlements face many problems, including:
- poor housing
- high crime rates
- lack of services (electricity, water and sanitation)
- unemployment
- underemployment (insufficient work, which is poorly paid).

With a rapidly growing urban population, conditions for many people are likely to get worse. Government policies through urban planning aim to improve their lives and address these problems.

Improving safety

Government actions include the following:
- Police and military patrols (including a Rapid Response Squad), an overnight ban on the use of motorcycles and the introduction of a toll-free emergency telephone number. This has had only patchy success, and criminality continues to increase.
- Private security guards in Victoria Island and Surulere. They are effective in reducing crime and popular with local residents. Some districts have street gates, which are closed at night.
- Increasing street lighting to make people feel safer at night.

Settlement upgrading

Upgrading (improving) policies usually focused on demolishing squatter settlements. These policies had little community involvement and were very unpopular. More recently, the government has been working with community groups to identify people's needs and consider strategies for improving the quality of people's lives.

With the support of the International Development Association, a seven-year upgrading programme (2006–13) focused on squatter settlements located on swampy ground:
- Over 1 million people have benefited from the upgrading of dilapidated roads and construction of new schools and health centres, as well as the provision of bathrooms and toilets and sinking boreholes to obtain water.
- Much of the work was carried out by members of the community who gained new skills. Land rights have also been granted to homeowners, encouraging further upgrading.

Other policies

Additional government policies include the following:
- An ambitious government plan seeks to use the vast area of Lagos Lagoon to house people in floating houses. In 2014 the Makoko Floating School was constructed to provide schooling for 60 children and act as a community centre.
- The Water Master Plan (2020) (see page 92).
- The Lagos State Government's 'Future Proofing Lagos – Energy Sector' project (see page 92).
- The Lagos State Water Regulatory Commission's efforts towards providing people with safe water (see page 94).

Figure 13.8 The floating slums of Lagos

Exam practice

1 With reference to a case study (Lagos), outline its national and international importance. (4 marks)
2 With reference to a case study (Lagos), describe the social opportunities associated with urban growth. (6 marks)
3 With reference to a case study (Lagos), describe how urban growth has created challenges. (6 marks)
4 With reference to a case study (Lagos), evaluate the management issues facing the city authorities. (9 marks)

ONLINE

13.3 Case study: Urban growth in Rio de Janeiro

This book uses Lagos in Nigeria and Rio de Janeiro in Brazil as case studies of major cities in an LIC or NEE. You only need to study **one** case study – one of these or another one that you have studied at school. If you studied a different LIC or NEE city, you might like to use these chapters (13.2 and 13.3) to guide you in making your own set of revision notes.

Rio de Janeiro – what is its location and importance?

With a population of 6.5 million people (together with another 12.5 million in the urban area), Rio de Janeiro (Rio) is the second most populous city in Brazil after São Paulo. It is located in the southeast of Brazil on the Atlantic coast (Figure 13.9), with most of the city built around Guanabara Bay.

Under Portuguese colonial rule, Rio was the capital of Brazil, becoming a major trading port. In 1960, the Brazilian government established inland Brasilia as the new capital city in order to encourage development in the interior of the country. Despite this, Rio remains one of Brazil's most important cities.

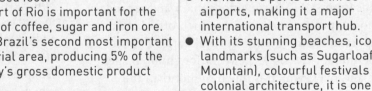

Now test yourself

Describe the location of Rio.

TESTED

Regional importance	National importance	International importance
● Rio is important in its provision of schools, universities and hospitals as well as many opportunities for employment, leisure and recreation. ● There is a thriving arts and cultural scene in Rio. ● The city is a major transport hub, with an international airport and important docks providing raw materials for local and regional industries, exporting products.	● Brazil's oil, mining and telecommunications companies have their HQ in Rio. ● Several of the country's universities and research institutions are based in Rio. ● Rio is a major manufacturing centre specialising in chemicals, pharmaceuticals, clothing and processed food. ● The port of Rio is important for the export of coffee, sugar and iron ore. ● Rio is Brazil's second most important industrial area, producing 5% of the country's gross domestic product (GDP). ● Brazil's major entertainment and media organisations are located in Rio.	● Rio hosted the 2016 Olympic and Paralympic Games and, in 2014, the football World Cup. ● The Statue of Christ the Redeemer is one of the world's most iconic landmarks, drawing tourists from all over the world. ● Rio is an international centre for industry and finance. ● Rio has five ports and three airports, making it a major international transport hub. ● With its stunning beaches, iconic landmarks (such as Sugarloaf Mountain), colourful festivals and colonial architecture, it is one of the most visited cities in the southern hemisphere.

Figure 13.9 Location of Rio de Janeiro

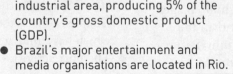

Figure 13.10 Rio

Revision activity

1 Create a spider diagram to summarise the regional, national and international importance of Rio.
2 Using Figure 13.10, describe the attractions for international tourists.

How has Rio grown?

Rio has a rapidly growing population, mainly due to migrants. Since 1950, the population of the city has trebled.

Essentially, population growth is due to the following:

- Migration – as Rio has grown into a major industrial, commercial and cultural centre, it has attracted large numbers of migrants from other parts of Brazil and abroad. One of the largest groups of migrants is Portuguese people – Rio is now considered to be the largest 'Portuguese city' outside Portugal itself! Rural–urban migration has been important as people have moved from the countryside (e.g. Amazonia) in search of better paid work and improved living conditions. Recently, Rio has attracted international migrants from South Korea and China wishing to establish businesses in the city. Today, Rio is a very mixed and cosmopolitan city.
- Natural increase – the high rate of migration has resulted in a relatively youthful population in Rio, which in turn has resulted in a high rate of natural increase, due to the relatively high birth rate.

Now test yourself

Explain how migration and natural increase have led to the rapid growth of Rio's population.

What opportunities are associated with the growth of Rio?

The growth and development of Rio has created a number of social opportunities and economic opportunities to improve people's quality of life.

Access to services (health and education)

- Health care – compared to remote, rural parts of Brazil, Rio offers better health care opportunities for new migrants, such as hospitals and health care centres. Vulnerable people, such as children and the elderly, have greater access to vaccinations and emergency care than if they were living in the countryside.
- Education – Rio has many primary and secondary schools (private and public), which offer better opportunities than exist in rural parts of Brazil. The literacy rate is 95 per cent for children aged ten and over, which is much higher than the national average. There are several universities offering tertiary education and training.

Access to resources (water supply and energy)

- Water supply – access to safe water has increased significantly in recent years, particularly in response to the city hosting major sporting events (the World Cup in 2014, and the Olympics and Paralympics in 2016). Over 90 per cent of the population now has access to mains water supply.
- Energy – many rural areas do not have lighting and power. Rio does provide better access, although supplies are subject to frequent power cuts, which cause problems for industry, water supply and other services. In poorer areas, many people tap into electricity lines illegally, which can lead to fires.

Improved economic development

Rio is one of Brazil's most important industrial cities, with its trading legacy and thriving port area. The expansion of its industrial areas has stimulated considerable economic growth:

- Improvements in roads, services (such as electricity and water) and the environment.
- The many manufacturing industries – chemicals, pharmaceuticals, clothing, furniture and food processing – attract migrant labour and economic investment.
- Sepetiba Bay in Rio is the largest steelworks in South America. New construction and supply industries have been attracted to the area, providing additional employment. This is called the multiplier effect. Similarly, the industrial area of Campo Grande in the West Zone has a large steelworks, which has stimulated growth in that area.
- The port activities themselves – dominated by the export of sugar, coffee and iron ore – provide employment opportunities.
- The Centro Zone in Rio is the main financial area, with the headquarters of Brazil's major oil company (Petrobas) and mining company (CVBB).
- The service sector (e.g. retailing, finance) has grown to serve people working in the industrial areas.

Revision activity

Construct a table to summarise the social opportunities associated with the growth of Rio.

Now test yourself

How can industrial areas be a stimulus for economic development?

What challenges are associated with the growth of Rio?

As the population of the city continues to grow, Rio faces many challenges to maintain and improve the quality of life of its people. The physical geography of Rio means that there are extremely high densities and limited opportunities for development.

Managing urban growth (slums and squatter settlements)

The lack of affordable property has forced millions of people to build their own homes – often no more than shacks – on land they do not own. These vast and sprawling squatter settlements – known locally as favelas – tend to be located on marginal land, such as alongside roads, on industrial wasteland or on the very steep slopes around the edges of the bay. Squatter settlements, also called slums, shanty towns and informal settlements, are found in cities throughout the developing world.

People living in Rio's squatter settlements face a number of challenges:

- Health – due to the extremely high population densities, disease can spread rapidly and infant mortality rates are high (50 per 1,000). Waste collection is limited, which increases the risk of disease.
- Landslides – following heavy rainfall, steep slopes are prone to landslides. In 2010, landslides swept away thousands of homes and over 200 people were killed.
- Services – in the poorest districts, over 10 per cent of people do not have access to piped water, 30 per cent have no electricity and 50 per cent lack proper sanitation. Drinking water may be contaminated or require collecting from far away. Electricity is often used illegally and subject to frequent power cuts.
- Building construction – most homes are constructed from waste materials (corrugated iron, wooden planks and even cardboard).
- Employment – unemployment and underemployment rates are high, while wages are low.
- Crime – in some favelas, crime rates are high, with violent crime and drugs being widespread problems.

Rocinha: Rio's largest favela

Rocinha has a population in excess of 100,000. Built on steep slopes above Copacabana and Ipanema (where many of its residents work), Rocinha has undergone considerable improvements:

- Ninety per cent of houses are constructed with bricks and have piped water, electricity and sanitation.
- There are schools, health centres and even a university.
- Many houses enjoy the relative luxury of a TV and fridge.
- There are shops and bars – and even a McDonald's restaurant!

Now test yourself

Suggest the main challenges facing the management of squatter settlements.

Providing clean water, sanitation and energy

The provision of clean water, sanitation and energy is a major issue in Rio.

Water supply

- Until recently, the vast majority of Rio's population did not have access to safe drinking water.
- Newer settlements, on the edge of the city that used to be part of the countryside, suffer the most. Here, residents obtain water from wells that are often polluted.
- One-third of piped water is lost through leaks and illegal tapping.
- Recent drought has increased the pressure on water supply as reservoirs have dried up (e.g. Paraibuna and Santa Branca).
- Between 1998 and 2014, the city authorities constructed seven new water treatment plants and laid 300 kilometres of new pipes to improve water supply.

Sanitation

- Many people have no access to flush toilets. Instead they use pit latrines, where sewage either drains into the soil (potentially polluting groundwater) or pours into open drains and rivers.
- Many of Rio's rivers and Guanabara Bay itself are heavily polluted with raw sewage – an estimated 200 tonnes pour into the bay each day.
- Providing people with proper sanitation is an enormous issue, particularly in the densely packed squatter settlements.

Energy

- Electricity supplies in Rio are overloaded and power cuts are frequent, affecting services (e.g. hospitals), industry, offices and residents.
- In squatter settlements, people sometimes illegally tap into electricity wires, but this can ignite electrical fires.
- Electricity supplies are being improved by the construction of the Simplicio Hydroelectric Complex on the Paraíba do Sul River, which will increase Rio's electricity supply by 30 per cent. The power complex became operational in 2013.

Providing access to services (health and education)

Health care

- Health care facilities are better than in most rural areas, but services are very patchy, with huge contrasts between wealthy and poor areas.
- For example, in Cidade de Deus in the West Zone of the city, only 60 per cent of pregnant females receive medical care and the infant mortality is a high 21 per 1,000. Life expectancy is just 45, compared with 80 in Barra de Tijuna in the South Zone, where medical care of pregnant females is 100 per cent.
- Improvements are being made – in the Santa Marta favela, health kits have been used to detect and treat diseases, resulting in a dramatic fall in infant mortality and an increase in life expectancy.

Education

- Education in Brazil is compulsory for all children aged six to fourteen.
- In Rio, only about 50 per cent of children continue their education beyond age fourteen as they mostly find work to support their families.
- School attendance is low, due to the lack of schools, a lack of teachers, the distance needed to travel to school and the lack of household money requiring children to work rather than attend school.
- The local authorities have introduced grants to help support children in schools and have encouraged local people to volunteer to work in schools.
- In Rocinha favela, a new private university has opened, and money is available for children to learn sports.

Reducing unemployment and crime

Unemployment

- Unemployment varies enormously between the rich and poor areas.
- In the favelas, unemployment rates can exceed 20 per cent.
- About one-third of the workforce work in the so-called informal sector, involving street vending, labouring, sewing, car washing or waste recycling. Such jobs are poorly paid, unregulated and often dangerous.

- While people working in this sector do not pay any taxes, they do not receive any insurance cover or unemployment benefit.
- The 'Schools for Tomorrow' programme is using education to teach practical skills to young people living in deprived areas to encourage them to seek formal employment. Similar skills-based courses are also available for adults.

Crime

- Crime rates are high, particularly those involving drugs, vandalism and theft.
- Violent crimes (assault and armed robbery) are common, with criminal gangs controlling drug trafficking in favelas.
- In 2013 Pacifying Police Units were established to address the problem of drug gangs and reclaim the favelas for the local community. The increased police presence has lowered crime rates and even led to tourism in some favelas.

Now test yourself   TESTED ☐

1 Outline the management issues associated with providing water and education in Rio.
2 What are the issues associated with reducing crime?

Managing environmental issues

Waste disposal

- Rio generates about 3.5 million tonnes of solid waste annually. Most waste is taken to a vast sanitary landfill site at Seropédica, 70 kilometres from the city centre.
- While full waste collection takes place in the city, collection in favelas is inadequate. Here, traditional collection vehicles have very limited access, due to the steep slopes and narrow alleys. Waste piles up, encouraging vermin such as rats and encouraging the spread of disease.
- Recycling is increasingly being encouraged and 'pickers' operate on the landfill sites to sort and recover materials for reuse.
- Landfill gas (LFG) is actively collected and harnessed as a source of energy, generating electricity and providing fuel for vehicles.
- A new biogas purification plant is being constructed at the recently closed Gramacho landfill site.
- Some organic waste (which makes up 50 per cent of household waste) is composted.

Pollution

- Pollution is a serious problem, causing premature death for thousands of people.
- For much of the time brown smog hangs over the city caused by emissions from the heavy traffic that clogs the streets, together with pollutants from the city's industries.
- Water pollution is a major problem, due to domestic sewage and industrial waste that pours into drains and rivers and eventually collects in Guanabara Bay.
- An estimated 200 tonnes of raw sewage pours into the bay each day, along with industrial waste and oil spills. The pollution threatens marine wildlife and the nearby tourist beaches such as Copacabana.
- Since 2004, twelve new sewage works have been built and 5 kilometres of sewage pipes installed to reduce the amount of raw sewage entering the bay.

Traffic congestion

- Rio is one of the most congested cities in South America, resulting in high levels of air pollution and wasted time for commuters and businesses.
- Traffic congestion is largely a reflection of the physical geography of the city, with its steep slopes, mountains and limited flat areas.
- To address the problem, the local authority is expanding the metro system, introducing toll roads in the city centre and making coastal routes one-way during peak times.
- Tunnels have been constructed to link different parts of the city.

Now test yourself   TESTED ☐

Outline the management strategies for reducing transport congestion in Rio.

How urban planning is improving the lives of the urban poor

The urban poor living in Rio's squatter settlements face many problems, such as:

- poor housing
- high crime rates
- lack of services (electricity, water and sanitation)
- unemployment
- underemployment (insufficient work that is poorly paid).

With a rapidly growing urban population, conditions for many people are likely to get worse. Government policies through urban planning aim to improve their lives and address these problems.

Favela Bairro Project

Since the 1980s the city authorities have encouraged the upgrading of favelas rather than adopting a policy of demolition and re-housing. Upgrading involves providing people with the materials and skills to improve their own houses while installing basic infrastructure, such as roads, electricity, water and sanitation.

The Favela Bairro Project is a 'site and service' project where the local authority provides land and basic services and residents build their own houses. The favela Complexo do Alemão in the north of the city is home to 26,000 people. Paved roads have been constructed along with access to water and sanitation. New health centres and schools have been built and a cable car transports workers to the commercial centre of Ipanema. Residents have access to credit to enable them to buy materials to improve their homes. A Pacifying Police Unit helps to reduce crime in the favela.

While the Favela Bairro Project has had some success, it is expensive, and the scale of the challenge is enormous as the favelas continue to grow. Infrastructure needs constant maintenance and people need training in constructional skills.

Exam practice

1 With reference to a case study (Rio), outline its national and international importance. (4 marks)
2 With reference to a case study (Rio), describe the social opportunities associated with urban growth. (6 marks)
3 With reference to a case study (Rio), describe how urban growth has created challenges. (6 marks)
4 With reference to a case study (Rio), evaluate the management issues facing the city authorities. (9 marks)

ONLINE

14 Urban growth in the UK

14.1 Urban growth in cities in the UK

How is the UK population distributed?

REVISED

The UK is one of the most urbanised countries in the world. About 82 per cent of the UK's population live in towns and cities. Figure 14.1 shows the distribution of population in the UK. It is a choropleth map showing population density (the number of people per square kilometre).

- The average population density is about 260 people per square kilometre.
- The Netherlands has a population density of over 400 people per square kilometre. Globally, the highest population density is Macau (China), with over 18,000 people per square kilometre!
- Population density varies considerably, from over 5,000 people per square kilometre in parts of London to fewer than 10 people per square kilometre in northern Scotland.
- The highest population density is in the London borough of Islington – 14,517 people per square kilometre!

Notice the following patterns:

- Population density is high across England, particularly in the south and east.
- The highest densities are focused on the major city regions (London, Manchester, Glasgow, etc.).
- The lowest densities are in the north and the west of the UK, coinciding with the uplands, especially in Wales and Scotland.

> **Exam tip**
>
> Remember that the term 'distribution' simply means describing where things are. The term 'pattern' can be used to imply a degree of regularity in the distribution – for example, you could talk about a linear or circular pattern. Be prepared to see both these words in exam questions.

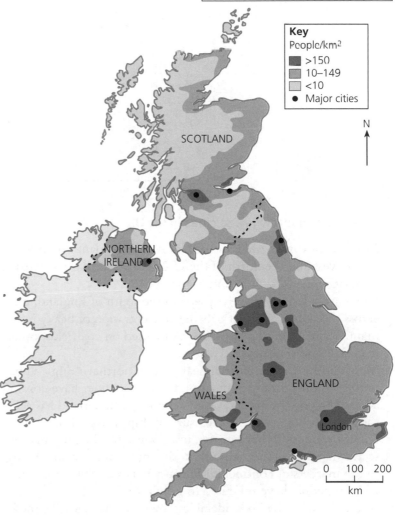

Key
People/km²
- ▇ >150
- ▨ 10–149
- ░ <10
- ● Major cities

N

SCOTLAND

NORTHERN IRELAND

WALES

ENGLAND

London

0   100   200
km

Figure 14.1 Population density in the UK

> **Now test yourself**
>
> Describe the distribution of population density in the UK.
>
> TESTED

Where are the UK's cities located?

Figure 14.2 shows the location of some of the UK's major towns and cities. You can see the following patterns:

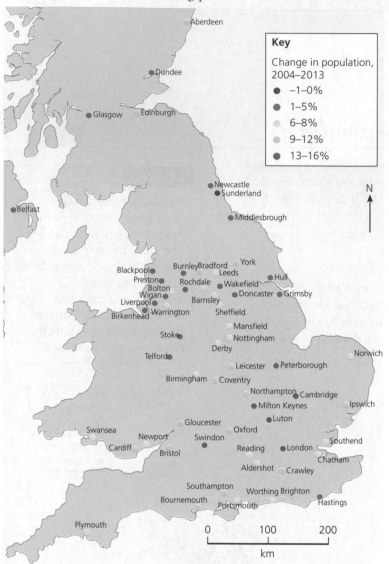

Figure 14.2 Major towns and cities in the UK

- Most major towns and cities are located in England, with relatively few in the west and the north of the UK. This is similar to the pattern of population density shown in Figure 14.1.
- Cities are relatively evenly spread in the south of England, but clustered across the north. This reflects the concentration of heavy industry and manufacturing in the past that was focused on coalfields, other raw materials and transport networks.
- Many cities in the industrial heartland of northern England have witnessed only a slow population increase as cities have continued to adjust to a post-heavy industrial economy. Sunderland has actually shown a decline, due to the closure of shipyards and coal mines.
- Cities in the south have grown in response to: the growing service sector, rapid economic growth in the south and east, and being close to the markets of continental Europe. Between 2004 and 2013, over a million people have migrated to London.
- Aberdeen witnessed considerable growth as the centre of the UK's oil and gas industry.

Case study: major city in the UK

You have to make a case study of one major city in the UK. There is a huge range of options and you will probably have studied your closest city. This makes good sense as you will be familiar with it, know some of the important locations and perhaps use it for conducting fieldwork.

Here is a template and guidance to help you make your own revision notes on your chosen city:

- Make sure that you address all the points listed in the guidance.
- Use some facts and figures where possible, but be selective: four to six specific bits of information is quite enough.
- Use simple sketches and revision diagrams (e.g. spider diagrams, bullets, tables) if they help you to revise.

London, Leeds and Bristol are some of the major cities in the UK, so it's likely you'll have come across information on them already.

Specification content	Revision notes guidance	Your revision notes
Location of the city	Focus here on the city's location in the UK. Where is it in relation to other places? Refer to both physical and human geography if possible.	
Importance of the city	Here you must focus on national (the UK as a whole) and international (global) aspects. A table would work well here. Try to identify three to five aspects for each, and be specific.	
Impacts of national and international migration on the growth and character of the city	Here you must focus on the impacts (effects) of people migrating from within the UK and from abroad. You should consider the impacts of migration on the growth of your city and on its characteristics (e.g. multi-culturalism – retailing, food, fashion, culture).	
Opportunities resulting from urban change – social and economic (cultural mix, recreation and entertainment, employment and integrated transport systems) and environmental (**urban greening**)	Opportunities are essentially positive outcomes. Make sure that you address each of the points listed – again, a table might work well. Don't forget to make some specific references to your chosen city – name localities and examples.	
Challenges associated with urban growth – social and economic (**social deprivation, inequalities** in housing, education, health and employment) and environmental (**dereliction**, building on **brownfield and greenfield sites**, waste disposal)	Challenges are essentially negative aspects that need to be addressed. Again, make sure that you address each of the aspects listed. Don't forget to make some specific references to your chosen city – name localities and examples. Look at the key terms to confirm your understanding.	
Impact of **urban sprawl** on the **rural–urban fringe** and the growth of commuter settlements	Urban sprawl – can have both positive and negative impacts (e.g. improved services but loss of habitats). Many villages have become commuter settlements – this may lead to economic growth but can also result in traffic congestion and a changing character of the settlement. Make sure you name your settlements. A sketch map could work well here.	

Exam practice

1 Study Figure 14.2 on page 104. Describe the distribution of major cities in the UK. (4 marks)
2 With reference to a major city in the UK, explain how urban change can create social and economic opportunities. (6 marks)
3 With reference to a major city in the UK, describe the environmental challenges associated with urban change. (6 marks)
4 With reference to a major city in the UK, assess the extent to which the advantages of urban sprawl and the growth of commuter settlements outweigh the disadvantages. (9 marks)

Exam tip

With Questions 2–4, stick rigidly to the wording of the question. For example, in Question 3 you should only write about **environmental** challenges.

ONLINE

14.2 Urban regeneration in the UK

You need to learn about **one example** of an **urban regeneration** project. You may well have studied a project in a city close to your school, so you can use the following as a template to summarise your notes. Remember that you will need to have some specific facts and figures about your chosen urban regeneration project. Here are two examples: the Olympic Park in London and Temple Quarter in Bristol.

Revision activity

Draw a spider diagram to identify the reasons why the area needed regeneration and the main features of the project.

Why do some areas need regeneration?

REVISED

Example

Olympic Park, London

What were the reasons for regeneration?

The Olympic Park is located in the Lower Lea valley (a tributary of the River Thames) in East London (Figure 14.3). A former industrial area, it was one of the poorest and most deprived parts of London. The area was characterised by low income housing, poor service provision, large tracts of derelict or underused land and polluted waterways. Warehousing, low-rise industrial premises, transport depots and railway sidings sprawled across the area.

The area was chosen for several reasons:
- Much of the area was either derelict, abandoned or occupied by relatively low-value land uses such as storage, warehousing, and so on.
- The deprived nature of the area meant that it was hoped that it would benefit hugely from the legacy of the Olympic and Paralympic Games in 2012.
- It is easily accessible from central London and elsewhere in the UK, particularly by train, and abroad.

What were the main features of the project?

Starting in 2007, urban regeneration included:
- the purchase of land under a single authority, the Olympic Delivery Authority (ODA)
- decontamination of land used for industrial purposes
- electricity cables buried beneath the ground to improve the appearance of the area
- construction of bridges over waterways to improve access
- landscaping to create natural habitats and improve the area's attractiveness.

When the Olympic Games opened in 2012, the area had been transformed. It now contains:
- excellent, modern sports arenas including the Olympic Stadium (now the home for West Ham United), the Aquatics Centre and cycling velodrome
- a landscaped park with tourist attractions and natural habitats
- the Athletes' Village, now converted into housing units for local people
- an improved aquatic environment, with footpaths and cycleways.

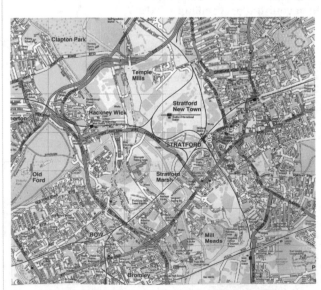

Figure 14.3 **Location of London's Olympic Park**

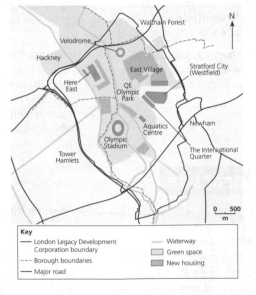

Figure 14.4 **The main features of the project**

Temple Quarter, Bristol

What were the reasons for regeneration?

Temple Quarter is an area of former industrial land close to Temple Meads railway station in the centre of Bristol (Figure 14.5).

The area became a focus for industrial development in the eighteenth century, with a thriving ironworks and gasworks ('town' gas used to be produced by processing coal). In the nineteenth century, improved waterways stimulated further industrial development, including potteries, glassworks and timber yards. People migrated into the area in search of work. Rows of back-to-back terraced houses were constructed to accommodate the workers. Further developments followed the construction of the railway station in 1841, opening up trade with the rest of the UK.

During the late twentieth century, decline set in. Industries closed down, land became derelict and the area became very run-down and polluted. By the 1990s, much of the area comprised unused railway lines, demolished slum housing and abandoned wasteland.

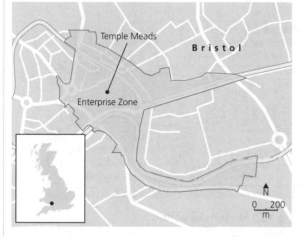

Figure 14.5 **Location map of Temple Quarter Enterprise Zone, Bristol**

What were the main features of the project?

The Temple Quarter Regeneration Project is one of the largest urban regeneration projects in the UK. Designated an Enterprise Zone, it has attracted public and private funding to transform the area into a thriving economic community.

- The 70-hectare Temple Quarter project was officially opened in 2012 and is expected to last for 25 years.
- The target is to create 4,000 new jobs by 2020 and 17,000 jobs by 2037. By 2015, 300 businesses had already been attracted into the area, creating over 2,000 jobs.
- Land has been cleared and decontaminated.
- Infrastructure is being improved: £21 million will be spent improving the vehicle, cycling and pedestrian access in the area, and a new bridge will be constructed over the River Avon; £11 million will be spent to provide superfast broadband for companies.
- New offices, houses and retail areas will be constructed and the area will be landscaped, with green areas, footpaths and cycleways.
- A new Bristol Arena will open in 2020, seating 4,000 people, for cultural and sporting events as well as conferences and exhibitions. The area will enjoy excellent access to other parts of the city and will have cafés, offices and flats.
- Brunel's Engine Shed will be transformed into a £1.7 million Innovation Centre, attracting start-up hi-tech firms as well as companies involved with the creative and low-carbon sector.

Revision activity

Draw a spider diagram to identify the reasons why the area needed regeneration and the main features of the project.

Exam practice

1 Define the term 'urban regeneration'. (2 marks)
2 With reference to an example of an urban regeneration project that you have studied, explain why the area needed regeneration. (6 marks)
3 With reference to an example of an urban regeneration project that you have studied, describe the features of the project. (6 marks)

ONLINE

Exam tip

Remember that this topic will be examined by reference to an example that you have studied, so make sure you have learned some specific facts and figures.

15 Urban sustainability

15.1 Urban sustainability

What are the features of sustainable urban living?

REVISED

Cities can have a significant impact on the environment by using resources, such as water, food and energy, and causing environmental damage through pollution. However, increasingly, cities are being planned for **sustainable urban living**. This involves cities becoming more self-sufficient, less demanding on resources and less damaging to the environment.

The following are various features of sustainable urban living.

Conserving water and energy

Water and energy conservation involves using less energy by using it more efficiently, and cutting down on waste, such as leaks from pipes. There are several ways of increasing conservation:

- Collecting (harvesting) rainwater and recycling water for use as 'grey water' in the house, at work and for community services (e.g. watering golf courses, parks, gardens). It is a waste of high-quality drinking water to use it to flush toilets and water gardens. In East Village (see page 106, Chapter 14) rainwater is collected and used for these things instead.
- Using biomass: in East Village, a combined heat and power system generates electricity using biomass. The heat warms water that is then piped directly to heat people's homes.
- Using green roofs to harvest rainwater and grow food – they also look attractive and create natural ecosystems.
- Increasing efficiency of devices such as washing machines, dishwashers, toilets (e.g. short-flush systems) and showers.
- Making greater use of insulation, double-glazing and heat-retaining building materials.
- Incorporating renewable energy systems such as solar panels on roofs.

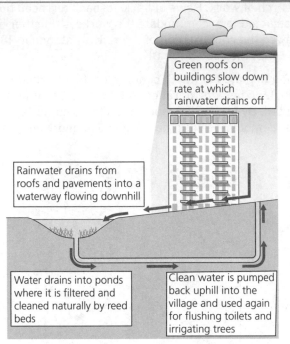

Green roofs on buildings slow down rate at which rainwater drains off

Rainwater drains from roofs and pavements into a waterway flowing downhill

Water drains into ponds where it is filtered and cleaned naturally by reed beds

Clean water is pumped back uphill into the village and used again for flushing toilets and irrigating trees

Figure 15.1 Water recycling

Recycling waste

Cities produce huge amounts of waste including sanitary waste, food waste (organic) and general rubbish (inorganic). Untreated waste can lead to significant health problems, causing the spread of waterborne disease such as cholera. People and wildlife can be poisoned by rotting waste, and vermin such as rats can quickly multiply. Options to increase waste recycling include the following:

- Reducing packaging and recycling materials, so that fewer of these end up as waste. The reduction in the use of plastic bags in the UK, since charging for their use began, has greatly reduced this type of waste.
- Using waste to create energy – e.g. biogas digesters convert organic food and garden waste into gas.
- Encouraging people to sort materials such as paper, cardboard, glass and plastic, and creating incentives for industry to do the same.
- Encouraging communities to grow food and compost organic waste.

Creating green space

Green spaces include public parks and private gardens. They act as the 'green lungs' of a city, absorbing carbon dioxide from the atmosphere and emitting oxygen. Urban greening absorbs and purifies water, helps reduce the flood risk, creates valuable natural ecosystems for wildlife and provides much-needed social and recreational space (walking, jogging, cycling, etc.).

Freiburg, Germany, is a good example of sustainable urban living.

Water and energy conservation	Waste recycling	Creating green spaces
• Waste water system collects rainwater for various 'grey water' uses. • Financial incentives encourage people to use water sparingly. • Green roofs are encouraged. • Permeable pavements and unpaved pathways encourage water to seep into the ground and reduce the flood risk. • Solar panels are used widely, e.g. Solar Business Park, railway station. Freiburg is a recognised centre for solar technology. • The city aims to be 100% powered by renewable energy by 2050.	• Burning waste provides energy for 28,000 homes. • Almost 90% of packing waste is recycled. • There are 350 collection points for recycling. • Financial rewards are given to people who compost green waste.	• Inner-city district of Vauban has been redeveloped to include green spaces between houses and green roofs for rainwater harvesting. • 40% of the city is forested. • Native trees are planted in the 600 hectares of public parks. • Much of the floodplain of the River Dreisam is kept free from development.

Urban transport strategies

Traffic congestion is a major issue facing cities, leading to increased levels of air pollution, affecting people's health. It also reduces economic efficiency, wasting time and money stuck in traffic jams. It can be reduced using the following measures:

- Developing an integrated transport system. In Freiburg, trams are widely used and integrated with other transport networks such as bus routes, cycleways and footpaths. There are over 30 kilometres of tramline and 400 kilometres of cycleways.
- Reducing car parking spaces or charging for cars to enter a city centre (e.g. Congestion Charge in London).
- Making public transport more attractive. In Singapore, automatic road pricing, high vehicle registration fees and high petrol prices deter people from using private cars, causing a 44 per cent reduction in traffic.
- Improving or adding to public transport. In Beijing, high parking fees and restrictions on car use have led to a 20 per cent drop in car use, and new metro lines and a rapid bus transit system have improved public transport provision.
- Enabling more cyclists on the road. In Bristol, a network of cycle routes encourages people to cycle rather than using private cars. There is also an integrated transport system involving railway (MetroWest), buses (MetroBus) and Park and Ride schemes.

> **Revision activity**
>
> Draw a summary spider diagram to outline the three features of sustainable urban living: water and energy conservation, waste recycling and creating green spaces.

> **Now test yourself**
>
> 1 What is 'sustainable urban living'?
> 2 How can water be conserved in urban areas?
> 3 What are the advantages of urban greening?
>
> TESTED ☐

Exam practice

1 What is meant by 'sustainable urban living'? (2 marks)
2 Evaluate the contribution of water and energy conservation to sustainable urban living. (9 marks)
3 Describe the transport strategies used to reduce traffic congestion. (6 marks)

ONLINE ☐

16 Economic development

16.1 Measuring development

The term **development** can be used to describe the progress of a country as it becomes more economically and technologically advanced. It can also be applied to improvements in people's quality of life, such as educational opportunities, increased incomes, human rights and healthy living conditions.

How is development measured across the world?

REVISED

The World Bank classifies the world into three broad groups according to measures of economic and social development:

- Low income countries (LICs) – roughly 30 of the world's poorest countries where most people have a poor quality of life with inadequate services and few opportunities. Most of these poorest countries are in Africa, with some in the Middle East.
- Newly emerging economies (NEEs) – these are countries such as Brazil, India and China, which are experiencing rapid economic growth and development, often based on industrial development. Incomes are rising and most people enjoy a reasonable standard of living.
- High income countries (HICs) – there are about 80 countries where most people enjoy a good standard of living based on relatively high levels of income. Most countries have efficient modern industries with a high proportion of people working in the service sector. HICs include most countries in northwest Europe, together with the USA, Japan, Australia and New Zealand.

Figure 16.1 is a development map of the world based upon the World Bank definitions.

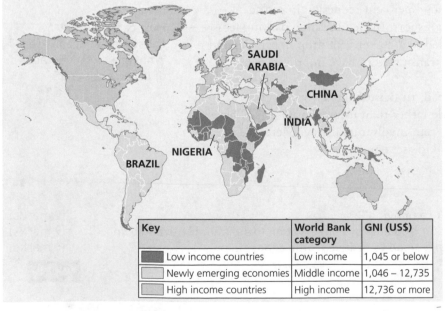

Key		World Bank category	GNI (US$)
▮	Low income countries	Low income	1,045 or below
▯	Newly emerging economies	Middle income	1,046 – 12,735
▯	High income countries	High income	12,736 or more

Figure 16.1 World Bank map of development

Now test yourself

1 Explain the meaning of the following acronyms: HICs, LICs and NEEs.
2 What is the meaning of the term 'development'?
3 Use Figure 16.1 to describe the global distribution of NEEs and HICs.

TESTED

What are the economic and social measures of development?

REVISED

There are several economic and social measures of development. While there are significant similarities between the global patterns produced, there are subtle differences, and some indicators tend to be more reliable than others.

It is important to remember that measures are averaged for a whole country. There will often be significant inequalities of wealth and social development **within** a country, particularly between major cities and remote rural areas. In fact, inequality is a good measure of the lack of development of a country!

Measure of development	Global variations	Limitations
Gross national income (GNI) – measured as GNI per capita (this means the total income divided by the number of people)	There are huge global variations – e.g. Norway (US$93,820), Somalia (US$150) – with most of the poorest countries being in Africa.	• These average figures can be misleading – a few very wealthy people in a country can distort the figures. • In poorer countries many people work in farming or in the informal sector, where their income is not taken into account by official GNI records. • Data about income is sensitive and people may not always be honest.
Birth rate – number of live births per 1,000 population	Generally, high birth rates are associated with poorer countries where child survival rates are low (due to poor health care, lack of safe water, poor diets and sanitation). Large families ensure a decent income for the family and provide support for ageing parents. The highest birth rates exceed 40 per 1,000 (Afghanistan, 44 per 1,000), with the lowest rates being about 10 per 1,000 (Finland, 11 per 1,000).	• Birth rates are quite a good measure of economic and social development. • Some countries (e.g. Cuba, 10 per 1,000) have a low birth rate even though most people are relatively poor. This is due to political decisions to focus investment in health care over other sectors. • Birth control policies can distort this as a measure of overall development (e.g. China, 12 per 1,000).
Death rate – number of deaths per 1,000 population	Death rates are relatively low throughout much of the world due to basic improvements in health care. Highest rates in Africa and parts of the Middle East (Afghanistan 18 per 1,000). Death rates can be moderately high in HICs due to the ageing population (Japan 9.9 per 1,000). Some of the lowest death rates are in NEEs where people are living longer but have yet to die of old age.	• Death rates are a poor measure of development. They can be high in some LICs due to poverty, but also high in HICs where there are many elderly people dying of old age.
Infant mortality – number of deaths of children aged less than one year of age per 1,000 population	Figures vary enormously, with the highest values in African countries (Angola, 96 per 1,000) and lowest values in HICs (Germany, 3 per 1,000).	• Recognised as a good measure of development as it reflects the levels of health care and service provision in a country. • In the poorest countries, not all deaths of children are reported, especially in remote areas – the true rates may be even higher.
Life expectancy – average number of years a person can be expected to live at birth	In HICs, life expectancy can be over 80 years. In NEEs, life expectancy is between 65 and 75. In LICs, life expectancy is typically in the 50s (Nigeria, 53).	• This is generally a good measure as it reflects health care and service provision. • Data is not always reliable, especially in LICs, and it can be slightly misleading in countries with very high rates of infant mortality – people surviving infancy may live longer than expected thereafter.

Measure of development	Global variations	Limitations
People per doctor – can also be expressed as doctors per 1,000 population	Huge variations exist between LICs and HICs – UK, 1 doctor per 350 people compared to Afghanistan, 1 doctor per 5,000 people.	● Increasingly people are seeking help and advice by using mobile phones – this is becoming popular in India and is not included in the data.
Literacy rates – percentage of people with basic reading and writing skills	Most HICs have literacy rates of 99%. In LICs, the figure can be below 50% (Afghanistan 38%).	● Another good indicator of development, though it can be hard to measure, especially in LICs due to the lack of monitoring. ● War zones and squatter settlements are difficult areas to measure literacy rates.
Access to safe water – percentage of people with access to safe mains water	In EU countries, all people should have access to safe water by law. Access in many LICs is poor (Angola, 34%).	● Data collection in LICs is not likely to be accurate and the official figures may underestimate the problem. ● People may technically have access but high costs may force them to use water that is not safe. ● Pipe leaks and natural disasters can deprive people of piped water.
Human Development Index (HDI) – composite measure using data on income, life expectancy and education to calculate an index from 0–1	Highest HDI values are in the HICs (Norway, 0.944) and lowest in the African LICs (Niger, 0.348).	● Developed by the United Nations, this is the most commonly used measure of development.

Now test yourself

TESTED ☐

1 What are the limitations of gross domestic income (GNI) as a measure of development?
2 Why is birth rate a better measure of development than death rate?
3 Why is the HDI one of the most widely used measures of development?

Exam tip

It is extremely likely that in an exam you will be given a map showing global development. It could be drawn using any of the development measures listed above.

What is the Demographic Transition Model?

REVISED ☐

The **Demographic Transition Model** (Figure 16.2) is a graph that plots changes in birth rates and death rates over time and shows how the total population grows in response. The Demographic Transition Model does not show the effects of migration on total population growth.

It is possible to link the model to development by identifying five stages, as in Figure 16.2.

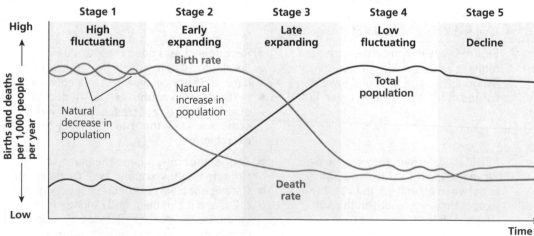

Figure 16.2 The Demographic Transition Model

Stage	Description	Links to development	Present-day country examples
1 High fluctuating	Birth and death rates are both high and fluctuating. They tend to cancel each other out, accounting for a stable but low population.	High birth rate reflects high infant mortality resulting from poor health care, diets, etc. High death rate results from disease and very poor health care and living conditions.	No countries are in Stage 1, though some small, isolated tropical rainforest tribal groups (e.g. in Indonesia) do have high birth and death rates.
2 Early expanding	Death rate starts to fall and then drops rapidly. Birth rate remains high. With an increasing natural rise, the total population starts to grow.	Improvements in basic health care and living conditions lower the infant mortality rate. This lowers the death rate as children are more likely to survive beyond infancy.	Afghanistan has a high birth rate (44 per 1,000) and falling death rate (18 per 1,000). About 80% of the population is involved in farming where children are needed to work on the land.
3 Late expanding	Death rate continues to fall before levelling off. Birth rate starts to fall rapidly. The total population continues to grow as birth rate exceeds death rate.	Further improvements in health care and living conditions (safe water, sanitation, diets, etc.) cause the death rate to fall further and then level off. Fewer children are needed to work the land and higher survival rates cause the birth rate to fall.	Many NEEs have these population characteristics, with a low and stable death rate and falling birth rate. Brazil's birth rate is 14.5 per 1,000 and death rate 6.6 per 1,000.
4 Low fluctuating	Both birth rate and death rate are low and fluctuating. As they cancel each other out, the total population growth slows and starts to level off.	High standards of health care, good living conditions and women choosing to study and follow careers reduce the birth rate. Economic conditions may cause the birth rate to fluctuate. Death rate remains mostly low and stable.	Most HICs are in this stage, enjoying relatively stable and reasonably sustainable populations. The USA has a low death rate (8 per 1,000) and low birth rate (13 per 1,000). Natural increase (5 per 1,000 or 0.5%) ensures that the population continues to grow slightly.
5 Natural decrease (decline)	Death rate remains constant but birth rate dips below, resulting in a natural decrease. This results in an ageing and declining population.	A fall in birth rate can result from an economic downturn or increasing numbers of women choosing to follow careers rather than having large families.	Some European countries, for example Germany, have experienced sustained periods of very low birth rates that have dipped below death rates. With ageing populations and fewer young people, this is an unsustainable population scenario.

Now test yourself

TESTED ☐

1 Explain the patterns of birth and death rate in Stage 1 of the Demographic Transition Model.
2 In which stage would you expect to find most NEEs, and why?
3 Why is the population unsustainable in Stage 5?

Exam practice

1 Use a labelled sketch to outline the main characteristics of the Demographic Transition Model. (4 marks)
2 Suggest the limitations of measures of economic and social development in classifying economic development and quality of life. (6 marks)
3 To what extent can the Demographic Transition Model (Figure 16.3) be linked to levels of economic and social development? (9 marks)

ONLINE ☐

Revision activity

Make a large copy of Figure 16.3. Draw text boxes below the diagram to summarise the links between population and development for each of the five stages.

16.2 Uneven development

Figure 16.1 (page 110) shows significant variations in levels of development across the world. This is known as the **development gap**. Both physical and human factors have contributed to uneven development.

What are the causes of uneven development?

REVISED

Physical causes

The physical geography of a country or a region can create several challenges for economic development:

- Weather and climate – extreme conditions, such as heavy rainfall, droughts, extreme heat or cold and vulnerability to tropical cyclones, create difficult conditions for economic development. Vast parts of central and western Africa experience limited and unreliable rainfall. Its population lives a very fragile life in a hostile environment. The Philippines and islands of the Caribbean are constantly ravaged by tropical storms. In 2016, over 1,000 people in Haiti were killed by Hurricane Matthew, just six years after 230,000 people were killed by a powerful earthquake.
- Relief – mountainous regions, for example countries such as Nepal, tend to be remote and have a poor infrastructure. They are also subject to extreme weather conditions.
- Landlocked countries – countries without a coastline lack the benefits of sea trade, which has led to the economic growth of most of the world's most developed nations. A coastline acts as an international border, providing huge opportunities for trading with other nations. Eight out of the fifteen lowest-ranking countries (according to the Human Development Index) are landlocked. Figure 16.4 shows the location of the world's landlocked developing countries.
- Tropical environments (hot and wet) – these are prone to pests and diseases, which can spread rapidly. Malaria, spread by mosquitoes, and waterborne diseases such as cholera can devastate communities and reduce people's ability to work.
- Water shortages – there are serious shortages of water, which is essential for life and for development, in some parts of the world, for example in parts of Africa and the Middle East.

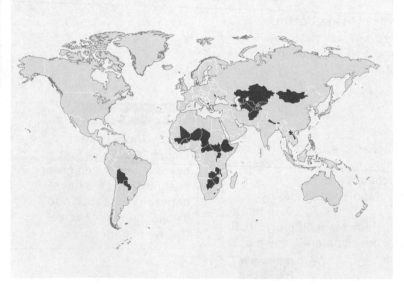

Figure 16.3 Landlocked developing countries (according to the United Nations)

Now test yourself

1 How can weather and climate affect economic development?
2 Study Figure 16.3.
   a Describe the distribution of landlocked developing countries.
   b Why does the lack of a coastline hinder economic development?

TESTED

Economic causes

The two main economic factors causing uneven development are poverty and trade:

● Poverty – the lack of money in a household, community or country slows development. It prevents improvements to living conditions, infrastructure and sanitation, education and training. Without the basics, developments in agriculture and industry will be extremely slow and an economy will simply fail to take off.

● **Trade** – trade between nations involves the import and export of goods and services. The vast majority of the world's trade involves the richer countries of Europe, Asia and North America. Most of the world's powerful international companies (TNCs) are based in the HICs. LICs have limited access to the markets. They have traditionally traded relatively low-value raw materials such as agricultural products or minerals, rather than higher-value processed goods. The value of these raw materials (commodities) has fluctuated wildly, causing great uncertainty and instability as countries strive to become developed (Figure 16.5). This trading imbalance has made HICs richer and increased the global development gap.

Many LICs were colonised by powerful trading nations such as the UK, France, Spain and Portugal. Much of Africa, South America (including the Caribbean) and Asia were brought under the economic and administrative control of powerful European empires. Many of these countries were exploited for their raw materials, and over 10 million people were exported from Africa to North America to work as slaves. It was during this colonial era that global development became uneven and populations were scattered. Most colonial countries became independent in the mid-twentieth century, e.g. India became independent from the UK in 1947 and Nigeria in 1960. However, since independence, many have been affected by power struggles and civil wars, and cope with the legacy of hundreds of years of exploitation. They face huge challenges, including poor infrastructure, lack of administrative experience and political instability.

> **Revision activity**
>
> Create a summary spider diagram to identify the main physical, economic and historical causes of uneven global development.

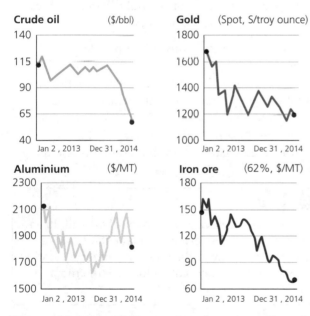
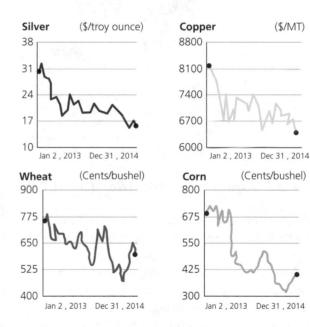

Figure 16.4 Recent fluctuations in commodity prices

Historical causes

Many HICs have experienced a long history of development based upon agricultural and industrial growth and international trading. They have therefore become highly developed and relatively wealthy. In recent decades, rapid industrialisation has taken place in countries such as China, Malaysia and South Korea (NEEs). Many other countries, however – the LICs – have yet to experience significant economic growth.

> **Now test yourself** TESTED ☐
>
> How has colonialism hindered economic development in many LICs?

What are the consequences of uneven development?

REVISED

Uneven economic development has led to disparities in wealth and health as well as high levels of international migration.

Disparities in wealth

We live in a very unequal world. There is a massive imbalance between the rich and the poor, with many people in LICs experiencing low levels of development, poor living conditions and poverty.

The highest levels of wealth are experienced by the most developed countries. Wealth is commonly indicated by the gross national income (GNI), one of the most common measures of economic development (see page 111). Figure 16.5 is a map of different income groups produced by the World Bank. It clearly shows the global disparities in wealth.

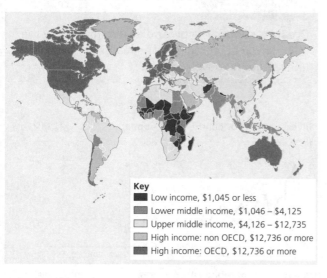

Key
- Low income, $1,045 or less
- Lower middle income, $1,046 – $4,125
- Upper middle income, $4,126 – $12,735
- High income: non OECD, $12,736 or more
- High income: OECD, $12,736 or more

Figure 16.5 Disparities in wealth (World Bank), 2016

- About 35 per cent of total wealth is held in North America by just 5 per cent of the world's population.
- Africa's share of global wealth is about 1 per cent (over 12 per cent of the world's population).
- China is one of the fastest-growing countries in terms of wealth – along with India, average personal wealth has quadrupled since 2000.

Disparities in health

Countries with a low level of development tend to experience poor health care. Most people in LICs have limited access to doctors, health clinics or hospitals. They experience relatively high rates of infant mortality and low life expectancy:

- In LICs, 40 per cent of deaths are in children under 15, compared to 1 per cent in HICs.
- In HICs, 70 per cent of deaths are people aged over 70, compared with 20 per cent in LICs.

In the most extreme cases, children in particular can suffer from malnutrition (a basic lack of food) or malnourishment (a lack of a balanced diet, i.e. lacking certain specific nutrients). Generally, however, disease is the direct cause of death:

- In LICs, malaria and tuberculosis account for one-third of all deaths.
- Malaria kills over half a million people a year, mostly in Africa. Here the environmental conditions create ideal breeding grounds for mosquitoes and the lack of preventative measures means that many people become infected and die.
- In HICs, chronic diseases such as cancer, heart disease and dementia are the main causes of death.

International migration

In 2015 some 14 million people migrated from poorer countries to seek a better life elsewhere. There are different types of international migrant:

- Economic migrant – a person who moves voluntarily to seek a better life abroad.
- Refugee – a person who has been forced to move, often due to conflict or natural disaster, and seeks sanctuary in a foreign country.

In recent years, many thousands of people have migrated into Europe from war-torn regions of Africa and the Middle East, such as Syria, Afghanistan and Libya. They often undertake dangerous journeys across the Mediterranean to reach Italy, Greece or the Canary Isles (Spain), and many have died on the way. In 2015, an estimated 1.1 million migrants entered Germany seeking a better life.

Exam tip

In Question 3, below, you need to refer to all three causes when considering 'to what extent'. Make sure that you answer the question, either by stating that you do think physical factors are the most important or that you do not.

Exam practice

1 Study Figure 16.4.
   a Describe the pattern of commodity prices shown by the graphs. (4 marks)
   b Suggest why a reliance on the export of commodities hinders economic development. (6 marks)
2 Explain the consequences of uneven development. (6 marks)
3 To what extent is the development gap the result of physical factors? (9 marks)

ONLINE

17 Strategies for reducing the development gap

17.1 Reducing the development gap

How can we reduce the development gap?

There are several strategies that can be used to reduce the development gap. Many of these support local development projects (e.g. agriculture, water and energy). These generate employment opportunities and increase incomes, which in turn create more wealth and help to reduce the development gap.

Strategy	Details
Investment	Countries, organisations (e.g. the World Bank) and TNCs invest in LICs to increase profits. Investments lead to improvements in infrastructure (e.g. airport construction), services (e.g. water and electricity), dams and reservoirs (for hydro-electric power (HEP) and irrigation) and industrial development. Over 2,000 Chinese companies have invested in Africa, e.g. HEP in Madagascar and the Tazara railway linking Tanzania and Zambia. Companies such as Google, IBM and Walmart have all invested in Africa. Investment can provide employment opportunities and increase incomes, thereby reducing the development gap.
Industrial development and tourism	Industrial development provides employment, increases individual wealth and results in improvements in education, health care and service provision. Industrialisation has promoted development in countries such as Brazil, Mexico and Malaysia (e.g. the development of the Proton car). Tourism provides a valuable source of foreign exchange and can lead to improvements in infrastructure (airports, roads, electricity and water), education and health care. Tourism creates employment opportunities and raises incomes.
Aid	Aid often takes the form of financial support offered by countries, international organisations (such as the EU) and charities. Short-term emergency aid is given in response to natural disasters. Long-term aid supports development projects, such as improving water supply, sanitation and energy provision. UK aid supports education projects in Pakistan.
Intermediate technology	Intermediate (sustainable) technology is appropriate in many countries for supporting local development projects involved with agriculture, water and health. For example, local labour and materials were used to construct a small dam to improve water supply at Adis Nifas in Ethiopia. Through improvements at the grassroots level, the development gap can be reduced.
Fairtrade	Fairtrade is an organisation that promotes fair wages for farmers in LICs. The international organisation guarantees the farmer a fair price and invests money in local community projects. In Uganda, coffee farmers have benefited from Fairtrade, processing their own coffee beans to increase the export value of the crop. This helps to reduce the development gap.
Debt relief	Many countries borrowed money in the 1970s and 1980s to invest in development projects. Some countries have fallen into serious debt, unable to pay back the loans or pay the high rates of interest. In 2006, the International Monetary Fund (IMF) agreed to cancel the debts of nineteen of the world's poorest countries. This money can now be used for development projects (e.g. road building and health care in Ghana), improving the lives of millions of people and helping to improve the development gap.
Microfinance loans	Small-scale financial support to help individuals or community groups to start small businesses. If successful, these businesses will create jobs and increase people's incomes. A good example is the Grameen Bank in Bangladesh. The bank lends money to women to buy a mobile phone so that they can check market prices when selling agricultural produce.

> **Exam tip**
>
> You need to learn about each of the strategies listed in the table above as they are all listed in the specification. Make sure you understand how each strategy helps to reduce the development gap – you must be prepared to show that you understand this link.

How the growth of tourism can reduce the development gap

You need to choose **one** example to illustrate how the growth of tourism can reduce the development gap. Here are two for you to choose from. If you have studied a different country, make similar revision notes on that country.

Country details	Reducing the development gap
Tunisia ● Attractive climate, with hot and sunny summers and mild winters ● Historic and cultural attractions (e.g. ancient city of Carthage) ● Mediterranean beaches ● Sahara desert ● Good air connections with Europe	Tunisia ● Tourism has created 370,000 jobs and boosted incomes, increasing money circulating in the economy. ● Local businesses have benefited from the development of coastal resorts (e.g. construction industry, shop owners, taxis). ● Agricultural sector has benefited by providing food for tourists. ● Government has invested money in the health service and in education. ● Literacy rates have increased, as has life expectancy.
Jamaica ● Popular destination in the Caribbean ● Excellent all-year tropical climate, with high temperatures throughout the year ● Stunning sandy beaches with opportunities for watersports, deep-sea fishing and scuba diving ● Rich cultural heritage (e.g. plantation houses) ● Excellent air communications and a popular destination for cruise liners ● Wildlife reserves, bird sanctuaries, botanical gardens	Jamaica ● In 2014 tourism contributed 24% of Jamaica's gross domestic product (expected to rise to over 30% by 2024). ● Tourism income exceeds US$2 billion each year. ● Tourism is the main source of employment – over 200,000 local people are employed in the sector (e.g. hotels, shops, agriculture). ● Increased incomes have benefited the retail sector and increased the amount of money circulating in the economy. ● Infrastructure has been improved to support tourism, e.g. new port facilities, roads and hotels. ● Many local people in the key tourist sites of Montego Bay and Ocho Rios have witnessed improvements in their quality of life, though there are still pockets of poverty. ● The environment has benefited by landscaping projects (e.g. Montego Bay) and the designation of nature parks (e.g. Negril Marine Nature Park).

Figure 17.1 Tunisia

Figure 17.2 Jamaica

Exam practice

1 Outline the role of aid in helping to reduce the development gap. (2 marks)
2 Describe how intermediate technology and microfinance can reduce the development gap. (6 marks)
3 'The growth of tourism can reduce the development gap.' Assess the validity of this statement with reference to an example that you have studied. (9 marks)

ONLINE

18 Economic development in the world

18.1 Case study: economic development in Nigeria

The specification requires you to study economic development in a low income country (LIC) or a newly emerging economy (NEE). This chapter focuses on Nigeria. If you have studied a different country, you could use the following headings to help you structure your own revision notes.

What is Nigeria's location and importance? REVISED

Nigeria is a country in West Africa bordered by Benin, Niger, Chad and Cameroon (Figure 18.1). Southern Nigeria borders the Gulf of Guinea, part of the Atlantic Ocean. Located just to the north of the Equator, Nigeria experiences a range of climates and natural environments, with tropical rainforests in the south and semi-desert in the north. It is approximately three times the size of the UK and has a population of around 180 million, also about three times that of the UK.

Nigeria is the most populous and economically powerful country in Africa. In recent decades economic growth, largely based on oil, has transformed the country from an LIC to an NEE.

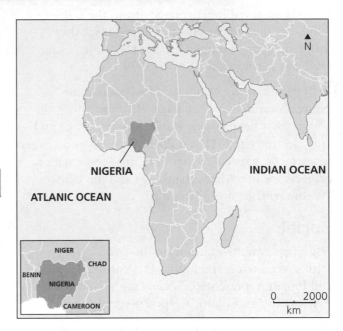

Figure 18.1 Location of Nigeria

Global importance	Regional importance
• Nigeria is the world's 21st largest economy and has experienced a rapid growth in gross domestic product. • Nigeria is ranked seventh in the world by size of population. • Nigeria is the world's twelfth-largest oil producer. • Lagos is a thriving 'world city', with a strong economic and financial base. • Nigeria plays an important peacekeeping role in world affairs.	• Nigeria is one of Africa's fastest-growing economies and has the highest gross national product (GNP) on the continent. • Nigeria has the third-largest manufacturing sector in Africa and the continent's highest population. • Nigeria has the highest farm output in Africa. It has the highest number of cattle. • Nigeria is generally seen as an indicator for the entire continent – if Nigeria thrives, Africa will thrive.

Now test yourself

List three facts each to describe Nigeria's global and regional importance. TESTED

What is the wider context?

REVISED

Political

In 1883, European 'superpowers' created the political map of Africa. For many years, Europeans exploited African resources and promoted slavery.

Nigeria achieved independence from the UK in 1960, but then experienced political instability as different factions fought for control, with a bitter civil war raging from 1967 to 1970.

In 1991 the newly built city of Abuja became Nigeria's capital city (see page 91). From 1999, the country has been largely stable, enjoying 'free and fair' elections in 2011 and 2015. This has encouraged investment, particularly from China (construction projects), South Africa (banking) and the USA (Walmart, IBM and Microsoft).

Social

Nigeria is a multi-cultural, multi-faith society, with several tribes, including the Yoruba, Hausa and Fulani, represented. Several faiths, including Christianity and Islam, are practised widely.

While social diversity is one of Nigeria's greatest strengths, it has led to some regional conflicts and power struggles. The rise of the Islamic fundamentalist group Boko Haram has caused conflict and hindered economic development.

Cultural

Nigeria's social diversity has created a rich and varied artistic culture, with thriving music, film and literary sectors. It has its own version of 'Bollywood', called 'Nollywood' – one of the largest film industries in the world. The Nigerian football team has won the African Cup of Nations on three occasions, and several players belong to Premier League clubs in the UK.

Environmental

Nigeria's global location places it within the Tropics:
- To the south, high rainfall promotes tropical rainforest. In this region, tree crops include cocoa, rubber and oil palm.
- Further north, as rainfall decreases, grassland (savanna) replaces trees. Here, people grow field crops such as millet, cotton and groundnuts (peanuts), and cattle graze on the savanna.
- An upland plateau region – the Jos Plateau – experiences cooler and wetter conditions (more favourable for farming) than the surrounding savanna.
- The far north has semi-desert conditions, with nomadic grazing of cattle.

How is Nigeria's industrial structure and economy balance changing?

REVISED

Nigeria is the largest economy in Africa, and one of the fastest-growing economies in the world. Despite this, almost 100 million people in Nigeria live on less than US$1 a day. Wealth tends to be focused in the south, around Lagos, with greater poverty in the north.

Nigeria's **industrial structure** has significantly changed. Nigeria's economy used to be dominated by the agricultural sector. Rapid industrialisation means that over 50 per cent of the country's gross domestic product (GDP) now comes from the manufacturing and service sectors (Figure 18.2):
- Employment in agriculture has fallen, partly due to increased mechanisation and rural–urban migration.
- Increased investment and political stability have led to a rapid rise in manufacturing.
- Along with industrialisation, there has been huge growth in the service sector (finance, communications and retail).

The oil and gas industry is hugely important to Nigeria's economy (see Figure 18.2). Discovered in the 1950s, oil and gas extracted from the Niger Delta fuelled Nigeria's industrial revolution and attracted massive foreign investment. However, fluctuating prices, and social and environmental issues in the delta region, have created economic turbulence.

Manufacturing is Nigeria's fastest-growing sector. It has a large, cheap labour force and a huge market. Many industries benefit from links with one another, for example petrochemicals, plastics and detergents.

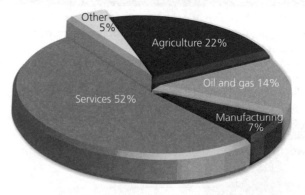

Figure 18.2 Nigeria's GDP by economic sector

TESTED

1 Describe how Nigeria's economic structure has changed.
2 To what extent can Nigeria be said to have a balanced economy?

How can manufacturing industry stimulate economic growth?

REVISED

The growth of manufacturing can stimulate economic growth in the following ways:

- Manufacturing industries often encourage the establishment and growth of linked industries, supplying raw materials or components to each other.
- Manufacturing stimulates growth of the service sector – finance, retailing and communications – leading to overall economic growth.
- Increased employment opportunities and higher wages increase consumer demand, which in turn increases the market, and leads to further growth and investment in industry.
- Increased employment results in higher taxes, which the government can use to expand the economy further and improve social conditions.
- A thriving manufacturing sector attracts foreign investment into the country.

How important are TNCs in Nigeria's economic development?

REVISED

Transnational Corporations (TNCs) have played an important role in Nigeria's recent economic growth. They can invest huge amounts of money and expertise while benefiting from tax incentives, cheap labour and large internal markets. TNCs located in Nigeria have good access to other African markets.

There are currently about 40 TNCs operating in Nigeria, most of which have their headquarters in Europe or the USA. Increasingly, Asian TNCs are investing in Nigeria.

There are several advantages and disadvantages of TNCs:

TNCs in Nigeria's oil and gas industry

The development of Nigeria's oil and gas industry depended on the investment and expertise of TNCs such as Royal Dutch Shell (UK, Netherlands), Chevron (USA), Exxon-Mobil (USA), Agip (Italy) and Total (France).

In the 1970s, TNCs invested heavily in oil and gas exploration, the construction of oil and gas platforms, laying of pipelines and the construction of oil and gas terminals. This investment has created employment opportunities (65,000 since the project began), raised incomes and contributed hugely to the Nigerian economy. Many Nigerian companies have benefited from the exploitation of oil and gas by winning contracts with the TNCs. However, the exploitation of Nigeria's oil and gas reserves by TNCs has been very controversial.

- Tankers transport oil to Europe and the USA where it is refined into petroleum products. The TNCs make most of their profits from refined oil.
- There have been many oil spills in the fragile deltaic environment, causing water pollution and damaging fisheries.
- Oil flares and toxic fumes have increased air pollution.
- Social unrest in the area has led to theft, sabotage and violent crime.

> **Revision activity**
>
> Construct a table or a spider diagram to identify the advantages and disadvantages of TNCs in Nigeria's oil and gas industry.

Advantages of TNCs	Disadvantages of TNCs
• Large companies provide employment and training of skills. • Modern technology is introduced. • Companies often invest in the local area, improving services (e.g. roads, electricity) and social amenities. • Local companies may benefit by supplying the TNCs. • TNCs have many international business links, helping industry to thrive. • The government benefits from export taxes, providing money that can be spent on improving education, health care and services.	• TNCs can exploit the low wage economy and avoid paying local taxes. • Working conditions may be poor, with fewer rules and regulations than exist in richer countries. • Environmental damage may be caused. • Higher-paid management jobs are often held by foreign nationals. • Most of the profit goes abroad rather than benefiting the host country. • Incentives used to attract TNCs could have been spent supporting Nigerian companies.

How is Nigeria's political and trading relationship with the rest of the world changing?

Political relationships

Nigeria's development from an LIC to an NEE has changed its global political position:

- Nigeria was originally part of the British Empire, so most political and trading links were with the UK and other members of the Empire.
- Since becoming independent in 1960, Nigeria has been part of the **Commonwealth**, maintaining strong links with the UK but also developing links elsewhere (particularly Africa, Asia and the USA).
- Nigeria plays a leading political role within Africa in terms of economic planning through the African Union and peacekeeping as part of the United Nations.
- Links with China are growing, as Nigeria benefits from increased investment, such as US$12 billion to construct a new 1,400-kilometre railway.

Trading relationships

Nigeria has strong trading relationships with Africa and the world:

- Imports: Nigeria's main imports are refined petroleum products from the **European Union** and the USA, cars from Brazil, mobile phones from China, as well as staple food crops such as rice and wheat. Most imports come from China, the USA and the EU.
- Exports: almost 50 per cent of Nigeria's exports are to the EU, and include crude oil, natural gas, rubber, cotton and cocoa. Most of Nigeria's crude oil is exported to India, China, Japan and South Korea. Approximately 30 per cent of Nigeria's cotton is exported to Australia and 15 per cent to Indonesia. Cocoa is exported for processing in Barbados.

Nigeria belongs to several trading groups: for example, the Economic Community of West Africa States (ECOWAS), a trading alliance with its headquarters in Abuja, and the Organisation of Petroleum Exporting Countries (OPEC). Figure 18.3 shows Nigeria's main trading partners.

1 How and why have Nigeria's political relationships changed since 1960?
2 Assess the importance of trading relationships with China.

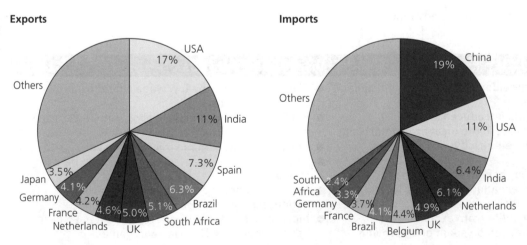

Exports

USA 17%
India 11%
Spain 7.3%
Brazil 6.3%
South Africa 5.1%
UK 5.0%
Netherlands 4.6%
France 4.2%
Germany 4.1%
Japan 3.5%
Others

Imports

China 19%
USA 11%
India 6.4%
Netherlands 6.1%
UK 4.9%
Belgium 4.4%
Brazil 4.1%
France 3.7%
Germany 3.3%
South Africa 2.4%
Others

Figure 18.3 Nigeria's main export and import partners

What has been the impact of international aid on Nigeria?

Aid involves the provision of support for people. It can take the form of emergency aid (e.g. food, water, shelter) following a natural disaster, or long-term development aid, aimed to improve people's quality of life (e.g. health clinics, water supply, schools). Figure 18.4 identifies the different types of **international aid**.

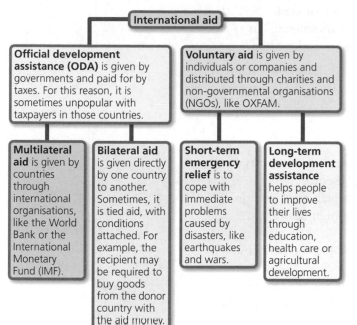

Figure 18.4 Types of international aid

Despite huge economic growth, poverty remains common in Nigeria (see page 93). Infant mortality rates are high and life expectancy low, especially in parts of northeast Nigeria.

Nigeria receives about 4 per cent of total aid given to African countries. Aid comes from organisations such as the International Development Agency/World Bank (medicines), the EU, UNICEF, and individual countries including the USA, the UK and Germany. The total amount of aid to Nigeria is about US$5,000 million per year.

Aid has benefited many people, particularly through community-based projects supported by small charities and non-government organisations (NGOs). Projects have included the following:
- The Aduwan Health Centre in northern Nigeria, supported by ActionAid and the World Bank, provides vaccinations and general health care, particularly for mothers and babies.
- Anti-mosquito nets are provided by the organisation 'Nets for Life'.
- Community Care in Nigeria, funded by USAAID, supports orphans and vulnerable children.
- Aid from the USA helps to educate and protect people against AIDS/HIV.

Despite good intentions, some aid money fails to get to the people who need it, due to corrupt individuals and corruption within the government.

Now test yourself

1 What are the different types of international aid?
2 How has international aid benefited poor people in Nigeria?

TESTED

What are the environmental impacts of economic development?

The rapid pace of economic development in Nigeria has had some harmful impacts on the environment.

Mining and oil extraction

This has resulted in serious incidents of pollution, particularly involving oil spills and fires in the Niger Delta, causing damage to aquatic ecosystems and toxic fumes being released into the atmosphere. Tin mining has polluted local water supplies and resulted in soil erosion.

In 2008/09, two massive oil spills devastated 20 square kilometres of natural swamps close to the town of Bodo in the Niger Delta. In 2015 Shell agreed to pay compensation to the community and to clean up the affected area.

Industrial development

The speed of industrial growth means that many large-scale industrial developments are unregulated and lack planning consent:
- In the major cities of Lagos and Kano, toxic chemicals are discharged into drains and open sewers, posing dangers to human health and natural ecosystems.
- Chimneys emit poisonous gases that can affect people and contribute to global warming.
- Deforestation is a major issue. Up to 80 per cent of Nigeria's forests have been destroyed. Burning releases carbon dioxide (a greenhouse gas) into the atmosphere and forest removal results in serious soil erosion.
- Waste disposal can poison rivers or the land, where it can contaminate groundwater supplies.

Urban growth

The rapid growth and outwards sprawl of urban centres such as Lagos have had a significant impact on the environment. Large areas of countryside have been lost, swallowed up by industrial developments or squatter settlements. The lack of sanitation and inadequate waste disposal cause land and water pollution, and the fumes caused by traffic congestion contribute to climate change.

> **Revision activity**
>
> Construct a summary spider diagram to identify the environmental impacts of economic development.

How does economic development affect people's quality of life?

Better paid jobs in manufacturing and services enable people to spend money improving homes or accessing health care and education.

Higher disposable incomes enable people to spend more money on recreational activities as well as food and clothing.

How does economic development affect people's quality of life?

Improved health care reduces infant mortality and increases life expectancy.

People are beginning to benefit from improvements in infrastructure (e.g. roads) and services (e.g. water, sanitation and electricity).

Improved living conditions enable people to perform better at school and at work.

Figure 18.5 How economic development affects quality of life

While many remain poor – particularly in rural areas in the north – a great many people have benefited from economic development. Nigeria's Human Development Index (HDI) has increased steadily since 2005, from below 0.47 to over 0.50 today. It has one of the fastest-growing rates of HDI in the world.

Since 1980:
- life expectancy has increased from 45.6 to 52.5
- access to safe water has increased from 46 per cent to 64 per cent
- expected years in schooling has increased from 6.7 to 9.0
- over 70 per cent of Nigerians now have mobile phone subscriptions and 38 per cent have internet access.

For the future, if the 60 per cent of Nigeria's population currently living in poverty are to benefit from economic development, certain challenges need to be met:
- The issue of individual and government corruption needs to be addressed to ensure that wealth reaches all people.
- Oil revenues need to stimulate growth across the economy and should be used to diversify Nigeria's industry.
- Environmental issues, such as soil erosion, desertification, malarial breeding grounds and oil spills, need to be addressed.
- Basic service provision (water, sanitation and electricity) needs to be a priority for all people.
- Ethnic and religious conflicts need to be addressed sensitively.

Only when the lives of ordinary people have been improved will Nigeria's economic development be considered a complete success.

> **Exam tip**
>
> In the exam, all questions requiring you to make reference to an LIC/NEE case study (e.g. Nigeria) will be phrased in a similar way to the following: 'With reference to a case study of an LIC/NEE country...' It is important that you learn your case study information thoroughly and be prepared to use this detail to support your answer.

> **Exam tip**
>
> Remember that with Questions 4 and 5 you will need to engage in some debate and discussion in order to achieve Level 3 marks. Make sure you focus on the question and use your case study to support the points you wish to make.

Exam practice

With reference to a case study of an LIC/NEE country:
1 Outline its global importance. (2 marks)
2 Describe the changes in the country's industrial structure. (4 marks)
3 Explain how economic development can have harmful impacts on the natural environment. (6 marks)
4 To what extent do the advantages of Transnational Corporations (TNCs) outweigh the disadvantages in promoting economic development? (9 marks)
5 Discuss the extent to which economic development has improved people's quality of life. (9 marks)

ONLINE

19 Changing UK economy

19.1 Economic change in the UK

At the height of the industrial revolution in the mid-nineteenth century, the UK's economy was dominated by manufacturing, including shipbuilding, and iron and steel (Figure 19.1). During the twentieth century, the UK's industrial structure changed:

● The primary sector (agriculture, mining and fisheries) declined, mainly due to the increased use of machinery.
● Since the 1960s, the manufacturing sector declined dramatically, also due to increased mechanisation and competition from abroad.
● Along with manufacturing, the service sector increased dramatically, due to the expansion in public services and the growth of financial services.
● Since the 1980s, the new knowledge–based (research and development) sector has become important.

In 2015, 78 per cent of the workforce was employed in the service sector, with just 10 per cent employed in manufacturing.

There are three main causes of economic change in the UK: deindustrialisation, globalisation and government policies.

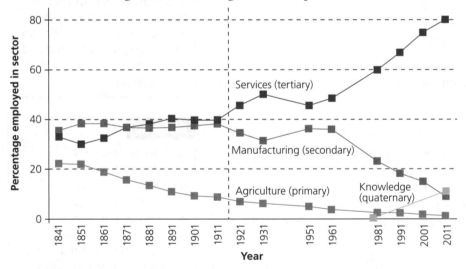

Figure 19.1 The changing industrial structure in England and Wales

What is deindustrialisation?

REVISED

Deindustrialisation is one of the most significant economic processes to have taken place in the UK. It has involved the decline in the UK's traditional heavy industries.

These industries were often based close to the raw materials (such as coal to manufacture iron, and steel for shipbuilding), for example in South Wales, Yorkshire, northeast England and Clydeside. These regions depended heavily on manufacturing. The decline of manufacturing and closure of coal mines from about the 1970s left a legacy of unemployment, low incomes and environmental dereliction in these regions.

What is globalisation?

REVISED

Globalisation has resulted from improvements in communications and technology, together with the development of trading groups such as the European Union.

The global economic landscape has been transformed by the growth of Transnational Corporations (TNCs) and the rapid economic growth experienced in Asia. The UK's place within this landscape has also changed to focus particularly on the service sector (e.g. finance, media, education) and the rapidly developing quaternary sector (e.g. research).

How have government policies addressed economic change?

REVISED

The government has an important role in shaping the UK's economy and responding to global trends. Since the Second World War, there have been three distinctive trends in government policy:

- 1945–79 – the government created state-run industries (e.g. British Steel Corporation) to support the UK's declining heavy industries and protect jobs. Outdated machinery and working practices led to factory closures, unemployment and considerable unrest during the 1970s.
- 1979–2010 – state-run industries were privatised, many heavy industries closed down and jobs were lost. Government and private sector investment resulted in a transformation of many former industrial areas (e.g. London Docklands) into financial centres (e.g. Canary Wharf), offices and retail parks as the service sector started to grow rapidly.
- 2010 onwards – the government has sought to rebalance the economy by encouraging investment in the high-tech manufacturing sector, for example aerospace and computer engineering. It has invested in transport infrastructure such as London's Crossrail and plans to develop high-speed rail connections (High Speed 2 – 'HS2') with the 'Northern Powerhouse'. Loans and other financial incentives are available to encourage small businesses to set up in the UK.

Revision activity

Create a large spider diagram summarising economic change in the UK. Draw a central box outlining the main changes that have taken place since the mid-twentieth century. Show the connections between this central box and the three main causal factors of deindustrialisation, globalisation and government policies.

Now test yourself

TESTED

1 Define the terms deindustrialisation and globalisation.
2 Outline the main government policies for each of the periods 1945–79, 1979–2010 and 2010 onwards.

Exam practice

1 What is meant by 'deindustrialisation'? (2 marks)
2 Explain how globalisation has led to economic change in the UK. (6 marks)
3 To what extent are government policies the most important driving forces in causing economic change in the UK? (9 marks)

ONLINE

Exam tip

In Question 3, make sure that you address the 'to what extent' by offering an opinion of your own, backed up by evidence.

19.2 The UK's post-industrial economy

Since the decline of the traditional heavy industrial sector in the 1970s, the UK has moved towards a **post-industrial economy**. This is characterised by the development of the following technology and sectors.

What effect has information technology had on the economy?

REVISED

The development of **information technologies** has transformed people's lives and economic development in the UK (see Figure 19.2).

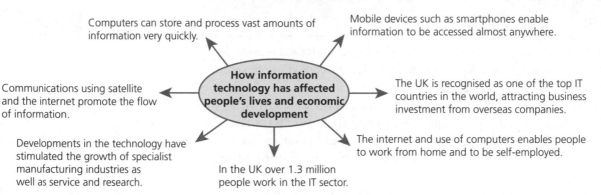

Computers can store and process vast amounts of information very quickly.

Mobile devices such as smartphones enable information to be accessed almost anywhere.

Communications using satellite and the internet promote the flow of information.

How information technology has affected people's lives and economic development

The UK is recognised as one of the top IT countries in the world, attracting business investment from overseas companies.

Developments in the technology have stimulated the growth of specialist manufacturing industries as well as service and research.

In the UK over 1.3 million people work in the IT sector.

The internet and use of computers enables people to work from home and to be self-employed.

Figure 19.2 How information technology has affected people's lives and economic development

What effect have service industries had on the economy?

REVISED

Service industries provide support rather than manufacturing products. The service industry is by far the largest sector in the UK, in terms both of employment and economic output. Finance, including banking, insurance and fund management, is one of the major growth areas.

The financial sector employs over 2 million people and contributes about 10 per cent of the UK's gross domestic product (GDP). The UK is recognised as the world's leading centres for financial management, with the City of London at its core.

What effect have developments in research had on the economy?

REVISED

Research and development is part of the UK's rapidly growing quaternary sector:

- It employs over 60,000 highly educated people and contributes £3 billion to the UK economy.
- Much research is linked to UK universities such as Oxford, Cambridge and Manchester and involves the biomedical, computer, environmental and aerospace sectors.
- Research is conducted by both government bodies (such as the National Health Service, the Ministry of Defence and the Environment Agency) and private organisations, such as pharmaceutical companies.
- It is likely to be one of the UK's fastest-growing industrial sectors in the future.

Now test yourself and Exam practice answers at **www.hoddereducation.co.uk/myrevisionnotes**

What effect have science and business parks had on the economy?

The growth of **science and business parks** has been an important recent trend in the UK's post-industrial economy.

Science parks

Science parks tap into research and employ recent graduates to apply academic knowledge to business innovation. Many businesses benefit from collaboration and share facilities, for example meeting rooms, restaurants and leisure centres.

Science parks are usually located on the edge of university cities such as Southampton, Oxford and Cambridge (see Figure 19.3), benefiting from good transport links and often enjoying attractive working environments. There are over 100 science parks in the UK, employing about 75,000 people.

- Located about 3 kilometres from Cambridge city centre.
- It enjoys excellent access to the A14 (to the Midlands) and the M11 (to London).
- Opened in 1970 by Trinity College. There are now several Cambridge colleges with links to businesses on the science parks.
- Many businesses employ graduates from the University of Cambridge.
- Staff benefit from an attractive working environment.
- The majority of the companies are involved with biomedical research, technical consulting or computer-telecommunications.
- Abcam ('antibodies Cambridge') is a biomedical company that is worth over £1 billion and employs 200 highly qualified graduates.

Figure 19.3 Cambridge Science Park

Business parks

Business parks usually involve a group of small businesses on a single plot of land. There are hundreds of business parks in the UK, usually located on the edges of towns and cities where land is relatively cheap and there are good road communications.

They can involve retailing and small-scale manufacturing, as well as research and development. In supplying goods and services, they often benefit from close association with each other (e.g. a printing company providing print materials for other companies in the business park).

Now test yourself

Define the following terms: information technologies, service industries, science parks and business parks.

TESTED

Revision activity

Draw a summary diagram to describe the characteristics of the UK's post-industrial economy.

Exam tip

When a question asks you to use information from an image, make sure your answer contains specific references to the image.

Exam practice

1 What is meant by a 'post-industrial economy'? (2 marks)
2 Describe the role of service industries in the UK's post-industrial economy. (4 marks)
3 Use the information in Figure 19.3 to explain why science parks such as the Cambridge Science Park form an important part of the UK's post-industrial economy. (6 marks)

ONLINE

19.3 What are the impacts of industry on the physical environment?

In the past, industrial growth had significant harmful impacts on the environment:

- Waste materials were often toxic, polluting the land and water supplies.
- Gas and soot emissions from burning coal polluted the air, resulting in smogs, mostly in London, in the 1950s.
- Landscapes in coal mining areas became transformed by ugly spoil heaps.

Modern industrial development is very conscious of its impact on the environment. These days, industry often builds on sustainable principles.

You need to study **one** UK example of a modern industry that is environmentally sustainable. This chapter considers two examples: the UK car industry and limestone quarrying.

Example

UK car industry

Each year the UK car industry manufactures over 1.5 million cars. Most of these are produced at seven giant factories owned by Transnational Corporations (TNCs), including Nissan, Honda and BMW.

Car manufacturing was not environmentally sustainable in the past. Engines were inefficient, producing high quantities of harmful pollutants as they burned petrol or diesel. The materials used to make cars were often toxic and difficult to recycle, and the production process was energy intensive. The situation is very different today.

Nissan car plant, Sunderland

Nissan employs 7,000 people in its car plant at Sunderland. Opened in 1986, it produces over 500,000 cars each year. It is highly efficient in the following ways:

- The car plant obtains 7 per cent of its energy from wind turbines.
- New car models are designed to be much more fuel efficient and have tighter restrictions on exhaust gas emissions.
- The Nissan 'Leaf' is an electric car. Other cars are 'hybrids', using a mixture of petrol and electricity.
- Cars are designed using materials that can be readily recycled, reducing waste going to landfill sites.

Example

Limestone quarrying

In the past, quarrying had some very harmful effects on the environment:
- Natural habitats were damaged or destroyed.
- Landscapes were transformed, often leaving ugly scars when quarries were abandoned.
- Water sources (rivers and aquifers) were polluted, particularly if toxic chemicals were used.
- There was extensive noise pollution and damage caused by huge lorries (which transported the quarried stone).
- Exhaust emissions caused atmospheric pollution.

Today there are very strict controls on quarrying to ensure that it is sustainable. There are strict regulations on blasting, removal of dust from roads and landscaping. All quarries have to be

restored after use, thereby minimising long-term environmental damage.

Torr Quarry, Somerset

Torr Quarry is a limestone quarry in the Mendip Hills, south of Bristol. It employs over 100 people and contributes around £15 million to the economy each year. Limestone chippings produced at the quarry are mostly used in the construction industry. The quarry is environmentally sustainable in the following ways:
- Chippings are transported by rail rather than road to minimise the environmental impact.
- There is regular monitoring of water quality, airborne emissions and noise.
- Some 80 hectares of the site have already been restored, with grass and trees.
- Future restoration will include the creation of lakes for wildlife and recreation.

Revision activity

Construct a summary diagram or table to identify the ways in which your chosen modern industry is environmentally sustainable.

Exam practice

Use an example to explain how modern industrial development can be environmentally sustainable. (6 marks)

ONLINE

Now test yourself and Exam practice answers at **www.hoddereducation.co.uk/myrevisionnotes**

19.4 Changes in the rural landscape

Rural landscapes in the UK are experiencing change. Some are experiencing population decline as younger people move away to seek jobs elsewhere. Other areas, particularly ones close to thriving towns and cities, are experiencing population growth and considerable social and economic changes.

The exam specification requires you to study social and economic changes in one area of population growth and one area of population decline.

What are the social and economic changes in an area of population growth?

REVISED

South Cambridgeshire

The city of Cambridge is one of the fastest-growing cities in the UK. The rural area of South Cambridgeshire surrounds the city. Its current population of about 150,000 is expected to rise to 175,000 by 2026 as people move to the countryside, and migrants from other areas – including from Europe – move into the city.

> **Exam practice**
>
> Explain the social and economic changes in the rural landscape in an area of population decline. (6 marks)
>
> ONLINE

Social changes	Economic changes
• Rising house prices (reflecting greater demand) and modern developments in villages can cause tensions with local people.	• Lack of affordable housing for local people.
• If a village has a high proportion of commuters it can become very dead during the day and lose its sense of community and identity.	• Some shops may be forced to close if commuters do not use their local village shops. Others may thrive if they offer services in the evenings and at weekends.
• Car-owning commuters do not need public transport and services may be reduced, affecting local people.	• Sale of agricultural land can reduce farm employment, which may lead to some local unemployment.
• Resentment may be felt towards migrants in the community.	• Fuel prices – and shop prices – tend to be higher than elsewhere in the area due to high demand.
	• The influx of poor migrants can put economic pressures on social services.

What are the social and economic changes in an area of population decline?

REVISED

Outer Hebrides

The Outer Hebrides are a remote group of islands off the far northwest coast of Scotland. The islands have experienced a 50 per cent decline in population since 1901, as (mainly young) people have moved to the mainland in search of better-paid jobs. The current population is about 27,000, most of whom live on the island of Lewis.

> **Exam tip**
>
> You could attempt the same exam practice question above for an area of **population growth**.

Social changes	Economic changes
• The out-migration of young people has resulted in an increasingly ageing population.	• Maintaining ferries and other services is costly – some post offices have had to close.
• An ageing population will require increasing amounts of care, which will have social and economic impacts.	• Traditional fishing for prawns and lobsters has declined, with just a few boats left.
• Fewer people of working age in the area could result in further decline in farming and fishing.	• Shellfish production has increased, mostly involving foreign boats.
• Fewer children could result in school closures.	• Tourism has become an important economic activity but the infrastructure is struggling to cope with the number of visitors.

19.5 Developments in infrastructure

There are now over 35 million cars on the road in the UK. The UK's infrastructure, particularly its transport networks, is under pressure as car ownership increases, and economic growth requires greater rail, port and airport capacity.

How are roads being developed?   REVISED

In 2014, the government launched a £15 billion 'Road Improvement Strategy' to improve road condition and capacity:

● Many stretches of busy motorway are being converted into 'smart motorways', enabling traffic flow to be controlled to reduce congestion.
● Over 100 new road schemes by 2020.
● Over 1,600 kilometres of new lanes will be added to busy motorways and major roads such as the heavily congested A303 in southwest England between Basingstoke (M3) and Exeter (M5).

These improvements will provide construction jobs for hundreds of people, and should boost local and regional economies throughout the country.

How is the rail network being developed?   REVISED

The government is keen to develop the UK's railways to ease congestion and help stimulate economic growth, particularly in northern England. Planned rail improvements include:

● London's Crossrail – due for completion in 2018, this £15 billion project, involving 32 kilometres of new lines bored beneath central London, will improve east–west connections across London and reduce commuting times
● electrification of the Trans-Pennine Railway – due for completion in 2020, this will improve connections between Manchester and York and will complete rail electrification between Liverpool and Newcastle
● High Speed 2 (HS2) – this £50 billion project involves constructing a high-speed rail line between London and Birmingham and then onwards to Manchester, Leeds and Sheffield. The route of the new line is controversial, particularly where it runs close to settlements or through highly valued countryside. Construction is due to start in 2017.

Figure 19.4 shows the routes of the new HS2 and Figure 19.5 lists some of the arguments for and against the new rail development.

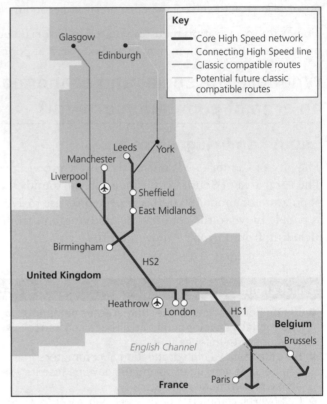

Figure 19.4 The proposed route for High Speed 2

What supporters day	What objectors say
● It will create thousands of jobs in the Midlands and northern England. ● It is estimated HS2 will help to generate £40 billion for the UK economy. ● It will increase the number of rail passengers and make transport more sustainable. ● It will also reduce the number of people who fly between UK cities. ● It will be a faster way to travel between cities. ● It will be carbon-neutral because it will reduce journeys that use other transport.	● It is more likely to create jobs in London and people will commute there instead. ● The cost of HS2 is estimated at £42 billion and it is difficult to predict how much money it will generate. ● Existing rail routes could be improved to increase the number of passengers. ● The number of people flying within the UK is already falling. ● People don't want to travel any faster. Intercity routes are already fast. ● It will increase carbon emissions because high-speed trains use more power.

Figure 19.5 Arguments for and against High Speed 2

What have been the developments in ports?

The UK has always been a trading nation, importing and exporting goods through its many thriving large ports such as London, Liverpool, Grimsby and Southampton (Figure 19.6). The UK's port industry is the largest in Europe, employing 120,000 people and handling 32 million passengers a year.

Figure 19.6 The location of the UK's largest ports

What have been the developments in airports?

Airports account for 3.6 per cent of the UK's gross domestic product (GDP) and are extremely important to the UK's economic development, providing thousands of jobs and boosting local economies. Over 2 million tonnes of freight pass through the UK's airports each year and 750,000 flights depart from the UK. The UK's airports handle close to 200 million passengers a year.

Heathrow expansion

Heathrow is the largest UK airport and one of the world's major hub airports, handling over 70 million passengers a year:
- In 2016 the government announced its intention to construct a third runway at Heathrow to ease congestion and enable expansion to take place.
- The highly controversial decision will have considerable impacts on local communities, including the demolition of properties in nearby villages. However, it should create thousands of jobs and boost the local economy.
- The new runway is estimated to cost over £18.6 billion and will be one of the most expensive infrastructure projects in the UK.
- There will be strict environmental requirements to reduce harmful emissions and limit aircraft noise.
- To reduce disruption to the nearby M25, the runway may be constructed on a ramp over the motorway!

The Heathrow expansion details are likely to change over the next few years, so spend a few moments updating your knowledge of this planned development!

Now test yourself

1 Describe **two** recent developments in road and rail improvements.
2 How can road and rail improvements lead to economic change in the UK?
3 Why are ports so important to the economy of the UK?

TESTED

Exam practice

1 Describe improvements and new developments in the UK's rail infrastructure.   (4 marks)
2 Explain how developments in transport infrastructure can change the UK's economy.   (6 marks)

ONLINE

Exam tip

Remember in Question 2 that you must link developments in transport infrastructure to change in the UK economy. Try to refer to a broad range of developments covering road, rail, ports and airports.

19.6 The north–south divide

What is the north–south divide?

The term **north–south divide** has been used to describe the cultural and economic disparities between the south of England – particularly London and the Southeast – and the rest of the UK:

● People living in the south tend to have higher incomes, longer life expectancy and a generally higher standard of living than those living in the north.
● High demand for housing in the south means that house prices are higher than in the north.
● In the north, unemployment rates are higher than in the south as areas continue to adjust to deindustrialisation.

These figures are averages. It's important to recognise that there are many pockets of poverty in the south – for example, some of the poorest boroughs in the UK are in London – and pockets of wealth in the north.

> **Now test yourself**
>
> What is meant by the north–south divide?
>
> TESTED

What has caused the north–south divide?

The main cause of the north–south divide is deindustrialisation, as traditional manufacturing industries, often based on raw materials such as coal, were largely based in the north – Yorkshire and Northeast England – as well as in South Wales (see page 126). As the northern economy declined, the economy of the south grew rapidly in response to the growth of the service sector and the dominance of London, particularly in financial services. This growth boosted average incomes and increased the value of property.

What strategies are being used to address the issue?

For several decades, the UK government and the European Union have attempted to restore the balance by investing in the north. Areas of the UK that are 'less economically advantaged' are recognised as having Assisted Area status (Figure 19.7). Financial assistance is available to support new businesses setting up in these areas.

Several schemes provide regional aid:
● The Regional Growth Fund supports projects in England that use private sector investment to create jobs.
● The Regional Selective Assistance encourages investment projects in Scotland, particularly in the Highlands and Islands.
● The Welsh Government Business Finance offers financial support to businesses and funds capital investment, job creation and research throughout Wales.
● Selective Financial Assistance provides support for investment in Northern Ireland.

> **Revision activity**
>
> Create a summary diagram or table identifying the strategies aimed at reducing regional differences in the UK.

> **Now test yourself**
>
> Outline **three** strategies aimed at reducing regional differences in the UK.
>
> TESTED

Other strategies include the following:

- Government incentive packages (reduced taxes, site development, and so on) have been used to attract Transnational Corporations (TNCs) to locate manufacturing plants, such as Mitsubishi near Edinburgh (1975) and Nissan in Washington, Tyne and Wear (1984).
- Planned transport improvements, such as HS2, rail electrification and port developments (see Figure 19.4, page 132).
- In 2015 the UK government launched the 'Northern Powerhouse' concept, encouraging industrial and infrastructural developments in northern England – cities such as Leeds, Sheffield and Manchester. The aim is to spread economic growth more evenly across the UK.
- Enterprise Zones – government incentives (such as discounted rates, provision of superfast broadband and simplified planning regulations) encourage new businesses to set up in deprived areas. Twenty-four Enterprise Zones have been designated since 2011.
- Local Enterprise Partnerships (LEPs) – these are voluntary partnerships between local authorities and businesses aimed at encouraging investment and boosting the local economy. For example, the Lancashire LEP is focusing development on the aerospace and technical engineering sector and aims to create 50,000 high-skilled jobs by 2023.

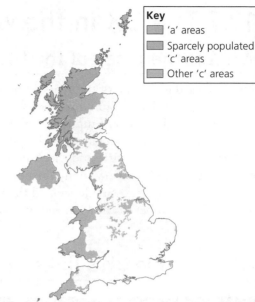

Key
- 'a' areas
- Sparsely populated 'c' areas
- Other 'c' areas

Figure 19.7 Assisted Areas in the UK

Type of Assisted Area	Relative level of aid	Proposed by	Regional characteristics
'a' area	High	European Commission	GNP per capita below 75% of the EU average
sparsely populated 'c' area	Medium	European Commission	Population density below 12.5 people per square kilometre
other 'c' areas	Low	UK government	Areas considered to be 'less advantaged'

Exam practice

1 Outline the evidence of a UK north–south divide. (2 marks)
2 Study Figure 19.7. Describe the distribution of the UK's Assisted Areas. (4 marks)
3 Explain the strategies used in an attempt to resolve regional differences in the UK. (6 marks)

ONLINE

Exam tip

In Question 2 be sure to explain how the strategies are intended to reduce the north–south divide. You must make the connection between the strategies and the reduction of regional differences.

19.7 The UK in the wider world

What is the place of the UK in the wider world?

REVISED

The UK used to be one of the world's most powerful political and trading nations:

● At its peak, the British Empire covered about a third of the world's land area, with colonies in Africa, the Asian Pacific and the Americas.

● In the twentieth century many former colonial countries gained independence and the UK became a member of the Commonwealth.

Today, the UK continues to have political, economic and cultural influence within organisations such as the G8, NATO and the UN Security Council. It remains one of the world's major economies and is a global transport and financial centre. The UK is also highly regarded for its fairness and tolerance, its highly developed legal system, its strong democratic principles and its rich cultural heritage.

What are the UK's links with the wider world?

REVISED

Trade	Culture	Transport	Electronic communications
● The UK trades with many countries by sea, air, road and rail (Channel Tunnel). ● The UK's main trading partners are the EU (particularly Germany, France and the Netherlands), the USA and China. Germany is the main source for imports and the USA the main destination for exports. ● The internet is increasingly important in the financial and creative sectors. ● Post-Brexit, the UK is likely to develop stronger links with countries outside the EU: for example, India, China and the USA.	● Cultural links include art, fashion, music, television and film. ● Television is one of the UK's major creative industries, worth over £1.25 billion a year. Programmes such as *Dr Who*, *Downton Abbey* and *Sherlock* are highly successful exports. ● Fashion, music and films are important exports, especially to the English-speaking world. ● Migrants to the UK have introduced their own cultural characteristics (foods, fashion, films and festivals).	● The UK's long trading heritage has resulted in the development of major ports and airports, such as Heathrow and Gatwick. ● There are links to mainland Europe via the Channel Tunnel, with fast rail services via the Eurostar and HS1. ● Ferries and cruise ships transport people to Europe and the rest of the world from ports such as Southampton and Dover.	● The internet is an increasingly important aspect of global communications – by 2014, 40% of the world's population had access to the internet (90% in the UK). ● The UK is an important hub for the global network of submarine cables linking Europe with the USA. ● Submarine cables are responsible for transmitting 99% of all internet traffic. ● The 'Arctic Fibre' project – due for completion in 2016 – involves 15,000 kilometres of cables linking London with Tokyo.

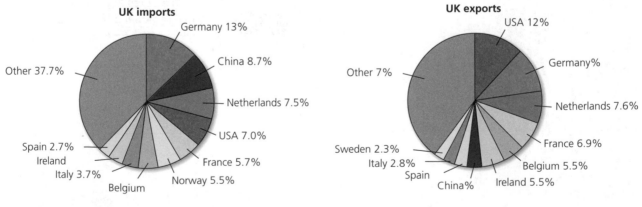

UK imports
Germany 13%
China 8.7%
Netherlands 7.5%
USA 7.0%
France 5.7%
Norway 5.5%
Belgium
Italy 3.7%
Ireland
Spain 2.7%
Other 37.7%

UK exports
USA 12%
Germany%
Netherlands 7.6%
France 6.9%
Belgium 5.5%
Ireland 5.5%
China%
Spain
Italy 2.8%
Sweden 2.3%
Other 7%

Figure 19.8 The UK's main trading partners

What are the UK's political and economic links with the European Union (EU)?

The UK joined the European Union (EU) in 1973. Today the EU comprises 28 countries, from the Baltic in the north to the Mediterranean in the south. It is one of the world's major trading blocs and exerts considerable political and economic influence.

In 2016, a UK referendum resulted in a majority decision to leave the EU ('Brexit') – this is due to happen in 2019. Until the UK officially leaves the EU in 2019 the UK remains part of the EU, benefiting from its support while obeying its laws and regulations. Membership of the EU has several effects on the UK:

- Goods, services, capital and labour can move freely between countries – sometimes called the 'free market'. The free movement of labour was a major factor in the 'Brexit' referendum vote.
- European funds, such as the European Structural Fund, support regional development in the UK.
- Hundreds of thousands of people from the poorer countries of Eastern Europe have entered the UK in search of higher wages. Many work in factories and in agriculture, willing to work long hours for relatively low wages.
- The Single Payment Scheme supports farmers, and benefits wildlife and the environment. In 2015, £18 million was used to support dairy farmers in England and Wales.
- Many EU laws and regulations affect working practices, product standards, safety, consumer rights and environmental guidelines. Some of these regulations can seem restrictive.
- The UK pays a considerable amount of money into the EU budget.

> **Now test yourself**
>
> Explain how the UK has developed important cultural links with the wider world.
>
> TESTED

What are the UK's political and economic links with the Commonwealth?

The UK maintains strong political and economic links with its former colonies through the Commonwealth, a voluntary organisation comprising over 50 countries. The aims of the Commonwealth are to provide support to individual countries and encourage co-operation between countries, all of which have a historic link. The heads of each country meet every two years to discuss issues of concern and work together to promote sustainable development.

Many people of British descent now live in Commonwealth countries, such as Canada, Australia and New Zealand. People have also migrated to the UK from Commonwealth countries such as India, Bangladesh, Nigeria and countries in the Caribbean. This movement of people has established strong cultural links between the Commonwealth countries and has also encouraged trade and business links.

There are important cultural and sporting links between Commonwealth countries. The Commonwealth Games, for example, is one of the world's major sporting events.

Exam practice

1 Describe the transport links between the UK and the wider world. (4 marks)
2 Suggest how the UK benefits economically from membership of the European Union and the Commonwealth. (6 marks)
3 To what extent does the UK play an important role in the wider world? (9 marks)

ONLINE

20 Resource management

20.1 The global distribution of food, energy and water

Why is food, water and energy important to economic and social wellbeing?

REVISED

Food, water and energy are essential to economic and social wellbeing. Where they are abundant, economies develop, societies thrive and people enjoy a good quality of life. Where they are scarce, economies are less likely to develop and people are less likely to enjoy a good quality of life. **Resource management** can have a significant impact on global and regional patterns of development.

Food

Food is an essential requirement for a healthy life. Food generates energy in the form of calories – you can see calorie data on the labels of most food and drink products.

Figure 20.1 shows the guidelines for average daily calories. Sports people, manual labourers and people living in extremely cold environments require a higher intake. In some parts of the world, mostly LICs, many people consume far fewer calories, which affects their wellbeing. At the other extreme, increasing numbers of people in HICs consume far too many calories and suffer from obesity, which also affects their wellbeing.

Category	Calories
Women	2,000
Men	2,500
Child, 5–10	1,800
Girl, 11–14	1,850
Boy, 11–14	2,200

Figure 20.1 Average daily calories

Water

Water is a basic requirement to sustain life. It is also needed to dispose of waste, grow and process food, and manufacture industrial products.

● In the UK, about 75 per cent of water is used in industry, 22 per cent for domestic purposes and 3 per cent in agriculture.
● Each person in the UK uses about 150 litres of water a day, of which just 4 per cent is used for drinking.

> **Revision activity**
>
> Choose an appropriate graphical method to represent the water uses in the UK.

Energy

The ability to create fire was one of the most significant advances in the development of the human species. Fire generates heat and energy, which can then be harnessed to do work. The introduction of machinery during the agricultural and industrial revolutions led to huge improvements in economic and social wellbeing.

Traditional forms of energy involve burning firewood and **fossil fuels** such as coal, oil and gas. Uranium is used in nuclear power plants to generate electricity. Increasingly, renewable energy sources such as water, wind and solar are being used to sustain our energy needs. Global energy consumption is heavily dominated by the use of fossil fuels.

> **Now test yourself** TESTED
>
> 1 How can too little and too much food affect people's wellbeing?
> 2 Why do sports people require a high energy intake of food?

> **Now test yourself** TESTED
>
> 1 What are fossil fuels?
> 2 Give three examples of renewable energy.
> 3 Why are renewable forms of energy being widely developed today?

How are global resources distributed?

Global resources are not evenly distributed across the world and this has greatly affected economic development and social wellbeing. See Chapters 22, 23 and 24 for detailed global maps of resource surplus and deficit. Here is a summary:

- Food – much of Europe, Asia and North and South America produce a food surplus, helped by a moderate climate, fertile soils and advanced technology. Most people in these regions can meet their daily food requirement. In Africa, however, physical conditions are more hostile and, together with low levels of technology and political instability, food production is less reliable. Many people do not receive their basic food requirement and suffer from undernourishment (basic lack of food) and undernutrition (lack of a balanced diet, i.e. lacking certain specific nutrients).
- Water – due to the climate, freshwater is also unevenly distributed across the world. Africa, together with parts of the Middle East, is prone to water shortages and drought. People have to spend huge amounts of time seeking water, and this can have a significant impact on economic development and social wellbeing.
- Energy – energy resources, particularly deposits of fossil fuels, are unevenly distributed. In the past, the availability of coal in Europe had a huge influence over early economic development and improving people's social wellbeing. In theory, renewable energy in the form of wind, solar and water is more evenly distributed. However, the high cost of development means that many poorer countries have been unable to tap into renewable sources.

To some extent food, water and energy are traded across the world to balance supply and demand. However, this has mostly involved HICs that can afford the high costs of imports. Many LICs, particularly in Africa, have not benefited greatly from the redistribution of resources.

Now test yourself

Describe the global inequalities in food.

How well off is the UK in terms of resources?

The UK is fortunate to have a resource surplus. This accounts for the UK's early and continued economic development and the relatively high level of wellbeing for most people.

- Food – the UK has a moderate climate strongly influenced by the Atlantic Ocean, with plenty of rainfall and mild temperatures. Benefiting from generally fertile soils, gentle relief and advanced technologies, the UK is one of the world's most efficient producers of food.
- Water – despite an imbalance of supply and demand within the UK (surplus: north and west, deficit: south and east), water supply is rarely an issue.
- Energy – the UK has large resources of fossil fuels (coal in the past, now oil and gas), several nuclear power plants (using imported uranium) and the potential for a range of renewable energies (wind, solar and hydro-electric power). In 2017, plans were announced to develop tidal barrages, for example in Swansea Bay.

Exam practice

1 Explain how food, water and energy contribute to economic and social wellbeing. [4 marks]
2 To what extent are economic and social wellbeing affected by global inequalities in resources? [6 marks]

21 Resources in the UK

21.1 Changing food demand in the UK

The UK's population is rising, which increases the UK's demand for food. In 2013, 47 per cent of the UK's food supply was imported to meet demand. The type of food we eat is also changing.

Why is there demand for high-value food exports? REVISED

It can still be cheaper for food to be grown in low income countries and transported to the UK, despite the increased **food miles**. High-value foods, such as Madagascan vanilla or specialist honey, can fetch higher retail prices than UK products.

Low income countries benefit from:
● wages for locals working in farming, packaging and transport

● taxes raised, which fund facilities such as schools and hospitals.

However, they have:
● less land for locals to grow their own food
● increased pressure on water supplies
● farmers exposed to chemical pesticides without protective clothing.

Why is there all-year demand for seasonal food? REVISED

Before supermarkets were commonplace, the majority of food eaten in the UK was seasonal food sourced in the UK. For example, strawberries only sold in summer and parsnips in winter.

Some foods cannot be grown in the UK's climate. The UK now demands greater choice in foods all year round, and these must be imported from other countries.

Why is there demand for organic produce? REVISED

Demand for **organic produce** has been rising since the early 1990s. Consumers are choosing meat, fruit and vegetables that reduce the negative impact on the environment and are healthier to eat. Instead of using chemical pesticides and fertilisers, organic produce is grown using:
● natural predators to control pests, e.g. ladybirds eat blackfly
● natural fertilisers, e.g. compost
● crops are rotated to maintain fertility
● animals are not fed drugs, e.g. hormones to increase growth.

Organic produce is more expensive because yields tend to be lower.

Figure 21.1 Santa strawberries

Now test yourself TESTED

1 What percentage of the UK's food was imported in 2013?
2 Identify three ways in which the demand for food is changing in the UK.
3 Describe two advantages for low income countries exporting food.
4 Describe two disadvantages for low income countries exporting food.
5 a Explain why Figure 21.1 is ironic.
  b What could have made demanding this food item possible in the UK?
6 Explain why consumers may choose organic produce.

What is the UK's carbon footprint?

In the UK, food travels over 30 billion kilometres every year. Food contributes at least 17 per cent of the UK's carbon dioxide emissions, of which 11 per cent is due to the transport of imported food. This increases the UK's **carbon footprint**.

Some foods grown in the UK have a larger carbon footprint than if the food were to be imported. This is because demanding produce such as tomatoes all year round requires heated greenhouses, whereas countries with warmer climates require no additional heating, thus reducing carbon dioxide emissions.

Planes generate emissions around 100 times greater than boats. Boats are slower so high-value perishable foods are generally transported by plane.

How does local sourcing reduce carbon emissions?

Local food sourcing reduces carbon emissions by:
- importing only foods unable to be grown in the UK
- eating seasonal UK produce
- eating locally produced food from local farmers' markets or farm shops
- home-growing food; a third of people now grow their own fruit and vegetables.

What is agribusiness?

Treating farming like a large industrial business increases food production. The size of farms is increased by removing hedgerows, combining small farms and an increased use of mechanisation and chemicals. **Agribusinesses** package and transport the food. However, employment in agriculture declines and there is a negative impact on the environment.

Figure 21.2 Riverford Farm Shop sign

Now test yourself

1 Identify evidence that shows the ways in which food contributes to the UK's carbon footprint.
2 Why might the carbon footprint not be reduced if the UK stopped importing food?
3 Which contributes least to the UK's carbon footprint? A plane or a boat?
4 What is meant by 'seasonal' produce?

Exam practice

1 Explain how local food sourcing reduces carbon emissions. (4 marks)
2 Explain how agribusiness increases food production. (4 marks)
3 Give a disadvantage of agribusiness. (1 mark)
4 Using the photograph of the farm shop sign and your own knowledge, explain the benefits of local food sourcing. (4 marks)

Revision activity

Write out the five key terms in this chapter, each on its own card. Write the key term on one side and the definition on the other (you can find the definitions on the AQA website or in your textbook). Use the cards to test yourself.

21.2 Changing water demand in the UK

How is demand for water changing?

The amount of water used by the average household in the UK has risen by 70 per cent since 1985. The Environment Agency predicts demand for water will rise 5 per cent by 2020. The growing demand for water is due to an increase in:

- UK population
- wealth, which means we have more water-intensive appliances (e.g. washing machines)
- showers/baths taken each week (especially power showers)
- demand for out-of-season food requiring watering in greenhouses
- industrial production
- leisure use (e.g. maintaining grass at golf courses).

Where are the areas of water deficit and surplus in the UK?

The UK receives enough rainfall to meet the UK's demand for water. Unfortunately the rainfall does not occur evenly throughout the UK or where there are most people. Annual rainfall is greatest in the west of the UK, while the east of the UK has a lower average rainfall. One-third of the UK's population lives in the southeast, the driest part of the UK.

Therefore the areas in the west of the UK with most rainfall have lower population densities so are areas of **water surplus**. The areas in the east of the UK have the least rainfall but higher population densities so are areas of **water deficit**. **Water stress** occurs when the water available is not sufficient to meet the needs of the population or is of poor quality.

> **Revision activity**
>
> Draw a rough sketch outline of the UK. Add arrows and labels to point out:
> a where rainfall is highest and lowest
> b where population densities are highest and lowest
> c where there is water surplus and water deficit.

Why is there a need for transfer to maintain supplies?

To match where the supplies of water are and where they are demanded in the UK requires water transfer. The UK government has considered a national water grid so water would flow through pipes from areas of surplus such as Wales to areas of deficit such as London. It has not been put into place due to a range of concerns as shown in the table.

Economic	● Enormous cost to install national water grid.
Social	● Local communities would need to be displaced.
Environmental	● Dams and reservoirs may disrupt ecology and block migrating species. ● Increased carbon emissions pumping water over long distances.

> **Now test yourself**
>
> 1 How is demand for water changing in the UK?
> 2 Why is water supply a problem in the southeast of the UK?
> 3 Give an example of where water in the UK is transferred from and to.
>

Parts of the UK have water transfers on a smaller scale. For example, the Kielder Dam in Northumberland pumps water into the North Tyne River which is then transferred to three other major rivers (Derwent, Wear and Tees) to supply Newcastle, Sunderland and Middlesbrough.

How good is the UK's water quality?

The Environment Agency manages **water quality** in the UK. UK waters are cleaner than they have ever been since before the industrial revolution, but only 27 per cent of water is classified as 'good status' under the EU Water Framework Directive.

Causes of water pollution in the UK

Rivers, lakes and coastal waters are polluted in a variety of ways, such as:
- chemical pesticides and fertilisers running from farm land
- hot water from cooling processes in industry pumped into rivers
- oil from boats and ships
- untreated waste (containing metals and chemicals) from industries
- oil, heavy metals from vehicle exhausts and road-gritting salt runoff from roads
- sewage containing bacteria pumped into rivers and sea
- inappropriate items (e.g. engine oil) put down drains.

Effects of water pollution in the UK

The effects of water pollution are both environmental and social.
- Pesticides kill aquatic wildlife.
- Fertilisers can speed up the growth of algae and lead to eutrophication. This means wildlife dies as there is insufficient oxygen in the water.
- Increased water temperatures can lead to death of wildlife.
- Toxic waste poisons wildlife, which can be transferred to humans if they eat contaminated shellfish or fish. This can lead to birth defects and even cancer.
- Drinking water can be poisoned.
- Microbacteria in sewage spreads infectious diseases in aquatic life and humans.
- Fishermen and the tourist industry, depending on clean water, suffer economic losses.

How are water quality and pollution levels managed?

- Legislation in the UK and EU means strict laws limit the amount and type of discharge factories and farms can put into rivers.
- Education campaigns inform the public about what is appropriate and what not to dispose of in sewage systems.
- Waste water treatment plants remove waste solids, bacteria, algae and chemicals so water is then safe for drinking.
- Investing in sewers and water mains reduces overflow of current sewers, spills and accidents.
- Pollution traps such as reed beds catch and filter out pollutants.
- Green roofs on buildings filter out pollutants in rainwater, reduce flooding, and combat climate change by absorbing carbon dioxide (CO_2) from the atmosphere.

Revision activity

1 Try to match some of the causes of water pollution in the UK to the effects of water pollution in the UK.
2 Try to match which management strategies would try to deal with each effect of water pollution.

Exam practice

1 Distinguish between water deficit and surplus. (1 mark)
2 Suggest two reasons why demand for water is increasing in the UK. (2 marks)
3 Explain why water transfer is required in the UK. (4 marks)
4 Explain how water quality and pollution is managed in the UK. (6 marks)

Now test yourself

1 Name four ways water is polluted in the UK.
2 Describe an environmental, an economic and a social effect of water pollution in the UK.
3 Name a method of managing water quality and pollution that can be undertaken by:
   a individuals
   b government
   c businesses and industry.

21.3 The UK's changing energy mix

How is the UK's energy demand changing?

REVISED

The UK consumes less energy than it did in 1970, even though there are 6.5 million more people living in the UK. The average household uses 12 per cent less energy. Heavy industry uses 60 per cent less due to its decline in the UK. Transport (both car and air travel) has increased its demand for energy since the 1970s due to its dramatic increase.

What is the UK's energy mix and how has it changed?

REVISED

- In 2015, the majority of the UK's **energy mix** is fossil fuels (coal 31 per cent, gas 25 per cent). They are non-renewable and emit carbon dioxide so contribute to climate change.
- Nuclear power provides just under a fifth of the UK's energy mix (19 per cent). Uranium produces heat in a nuclear reactor. Nuclear is non-renewable, but does not contribute to global warming.
- Renewable energy sources, such as **wind energy** and **solar energy**, provide just over a fifth of the UK's energy mix (22 per cent). They do not contribute to climate change.

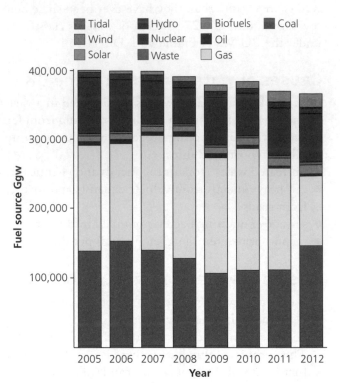

Figure 21.3 Energy production by fuel source, 2005–12

Reliance on fossil fuels

Until recently the UK had produced enough energy to power homes and industry with large reserves of oil and gas. These reserves have declined in the UK, so there is an increasing reliance on imported fossil fuels. Supplies can still be exploited in less accessible areas, but this is costly. In 2011 coal increased as older power stations worked to full capacity, as they were soon to be closed due to European Union regulations on emissions.

Nuclear energy has fallen slightly since the 1990s. Existing nuclear power stations are to close by 2023. New generation plants should be built and working by 2025.

Growing significance of renewables

Renewable energies, such as wind, are growing in significance, but are still only a small percentage of energy produced. Renewables are encouraged by the British government to meet targets to reduce carbon emissions and reliance on imports.

Revision activity

Draw a pie chart to show the UK's energy mix in 2015. (Label the remaining 3 per cent of energy sources of the pie chart as 'Other'.)

Now test yourself

TESTED

1 Why is it surprising that the UK consumes less energy than in the 1970s?
2 What sector demands more energy since the 1970s?
3 Describe the UK's energy mix.
4 What has happened to the UK's supplies of oil and gas?
5 How has the UK's reliance on renewable energy changed?

What are the issues with exploiting energy sources?

Energy exploitation of new or existing sources of energy has to happen in order to gain energy security. Exploitation raises several challenges and issues. There are economic costs as well as environmental impacts.

Non-renewable energy sources

Energy source	Economic issues	Environmental issues
Fossil fuels	• Non-renewable so unsustainable. Eventually the economic cost will be too high or they will run out. • Miners often suffer job-related diseases, which incur costs to the health service. • Costs increase to deal with effects of climate change and adaptation to it.	• Carbon dioxide is released, which contributes to climate change and acid rain. • Oil spills can devastate wildlife and sea life. • There is visual pollution from coal waste heaps and unsightly opencast coal mines. • Fracking for shale gas can cause groundwater pollution and earthquakes.
Nuclear	• Nuclear plants expensive to build and decommission, but the raw material (uranium) is cheap as such small amounts are used. • Cost to transport and store nuclear waste is high.	• Waste remains radioactive for over 100 years and has to be stored safely to avoid contamination. • Despite a good safety record, nuclear accidents release radiation into the atmosphere, which has long-term impacts on wildlife and people.

Renewable energy sources

Economic issues	Environmental issues
• Renewable energy has high set-up costs such as wind turbines, solar farms, hydro-electric dams and tidal power stations. Costs rise further in remote areas. • Biomass means land not used for food production may increase the cost of food. • Tourism declines as environments lose their visual appeal; results in job and income loss. • Low profitability is a concern.	• Many renewables are considered ugly and ruin the views in both the countryside and coast. • Wind turbines can affect bird migration and bat life. • Hydro-electric dams flood land upstream of the dam, changing the landscape and wildlife. Water held behind the dam changes temperature, affecting the ecology. Sediment is trapped by the dam. • Biomass reduces biodiversity as only one crop grown, e.g. sugar cane. • Geothermal energy is limited to tectonically active countries such as the USA and Iceland.

Now test yourself

1 Name the three fossil fuels.
2 State two economic challenges with fossil fuels.
3 State two environmental challenges with fossil fuels.
4 Give three examples of renewable energy.
5 State two economic challenges of renewable energy.
6 State two environmental challenges of renewable energy.

TESTED

Exam practice

1 Define 'energy exploitation'. (1 mark)
2 State an economic and an environmental issue with the production of nuclear energy. (2 marks)
3 'Fossil fuels have greater environmental challenges than renewable energy.' To what extent do you agree? (6 marks)

ONLINE

22 Food

22.1 Food: supply and demand

What are the global patterns of food supply?

REVISED

The world produces enough food for everyone on the planet; however, it is not distributed evenly. The countries with the highest levels of **food insecurity** are predominantly in Africa, the Middle East and parts of Asia (see Figure 22.1). Much of the rest of the world (HICs) enjoy **food security**.

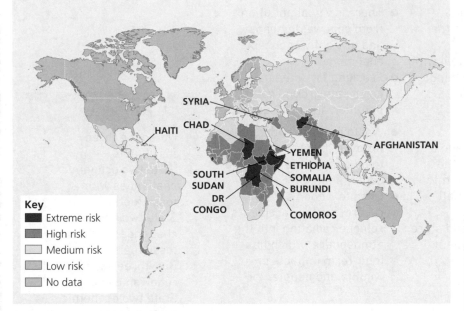

Figure 22.1 Global food security risk

On average people need about 2,500 calories a day to live a healthy life.
- Calorie intake is very uneven across the world.
- The highest calorie intake is in North America and parts of Europe (obesity is a growing problem).
- The lowest calorie values are found in Africa together with isolated countries in the Middle East and Asia. Many countries suffer serious food shortages leading to malnourishment (a basic lack of food) and **undernutrition** (a lack of a balanced diet, i.e. lacking certain specific nutrients). In the most serious cases, **famine** and starvation can result.

There are huge variations in calorie intake within individual countries, both rich and poor.

Revision activity

Look at Figure 22.1 and make a list of the main features of the map. Where are the areas of greatest food security? Which areas are at greatest risk from food insecurity?

Now test yourself

Define the following terms: food security, food insecurity, malnourishment and undernutrition.

TESTED

Why is food consumption increasing?

REVISED

Economic development

Rapid economic development, such as in China and India, increases the demand for food:

- As people become richer, they can afford to buy a greater quantity and variety of food. Richer people tend to eat more than poorer people, so as a country develops the demand for food (and its consumption) increases.
- Diets often change, with an increase in meat consumption. In China the percentage of meat in diets rose from 6 per cent in 1981 to 17 per cent in 2011. Livestock are fed cereals to produce meat for people to eat, whereas it used to be people who mostly ate the cereals.
- Demand for convenience and highly processed foods increases with economic development, as people have less time to grow and prepare their own food.

Population growth

Population growth increases the pressure on food production and supply. The world's population has grown rapidly since 1950 to its current level of about 7.5 billion. Almost all of this growth has been in LICs and NEEs where food production is low and many people suffer from food insecurity.

The fastest population growth is in Africa (2.51 per cent per year). This is the continent suffering most from food insecurity and low calorie intake. Feeding the people of Africa will be a huge challenge in the future.

Now test yourself

Outline how economic development and population growth have caused an increase in food consumption.

TESTED

What factors affect food supply?

REVISED

Several factors affect food supply, many of which are interconnected. They mostly affect African LICs.

Factor	Effects of food supply
Climate	Droughts, floods and climate change can have catastrophic effects on food production and distribution. • In sub-Saharan Africa, many farmers are dependent on seasonal rains which sometimes fail. Recent droughts in Ethiopia and Somalia have caused severe food shortages and migration. Droughts can lead to desertification and salinisation. • Serious flooding results from tropical storms which can devastate crops, e.g. Haiti and Fiji (2016). • Patterns of rainfall appear to be changing as a result of climate change. Some areas of the world may produce more food whereas others may suffer from more frequent droughts and floods.
Technology	Food supply and distribution in LICs can be affected by the lack of: • farm machinery (low yields) • irrigation • storage facilities • transport infrastructure (to distribute food) • processing facilities (to preserve food).
Pests and diseases	Many tropical regions suffer from pests and diseases and they often lack the money to protect crops and livestock: • Locusts can devastate food crops. • Cattle can be infected by airborne bacteria, causing bovine pleuropneumonia, a serious lung disease. • In poor societies, disease can reduce people's capacity to be productive.

Factor	Effects of food supply
Water stress	This is a serious issue in many LICs where climate change is expected to cause more damage: ● The lack of water security, together with drought, reduces food production. ● HICs can afford expensive water transfer schemes or irrigation projects, whereas LICs cannot.
Conflict	Recent conflicts in Somalia, Syria and South Sudan have led to food shortages. Conflicts often reduce food production and supply: ● War disrupts the distribution of food. ● Farmland may be mined. ● People are forced to move away; land (homes and farms) is abandoned. ● Water supplies may become polluted. ● Food aid may be restricted in areas of military conflict.
Poverty	Many farmers in LICs cannot afford high-quality seeds, fertiliser or mechanisation that would enable them to produce more food. They may also suffer from malnourishment (a basic lack of food) or undernutrition (a lack of a balanced diet, i.e. lacking certain specific nutrients), reducing their ability to work.

What are the impacts of food insecurity?

Famine and undernutrition

Famine can lead to malnourishment, undernutrition and weakened immune systems. The UN estimates that over 800 million people suffer from chronic malnourishment, almost all of whom are in LICs.

Soil erosion

Overcultivation and overgrazing together with lack of rainfall can lead to serious soil erosion, particularly in semi-arid regions. This reduces soil fertility and limits food production.

The impacts of food insecurity

Rising prices

When food supply falls, prices rise. Poor grain harvests in Russia, Australia and Pakistan in 2010 led to grain shortages and price rises across the world. If prices rise in local markets, the poorest people cannot afford a balanced diet so may suffer from undernutrition.

Social unrest

Food shortages can lead to rioting and social unrest. Between 2008 and 2011, 60 riots around the world were linked to food shortages, especially in North Africa and the Middle East.

Figure 22.2 The impacts of food insecurity

Revision activity

Draw a spider revision diagram to summarise the main causes of food insecurity.

Now test yourself

Explain how climate and conflict can affect food supply.

TESTED

Exam practice

1 Describe the pattern of food security shown in Figure 22.1. (4 marks)
2 Explain how food insecurity can lead to famine and undernutrition. (6 marks)
3 To what extent are physical factors more important than human factors in affecting food supply? (9 marks)

ONLINE

Exam tip

In Question 3 don't forget to address the 'to what extent' aspect of the question. This requires you to make a judgement and offer supporting evidence.

22.2 Food: increasing food supply

What strategies can be used to increase food supply?

There are several strategies, both in HICs and in LICs/NEEs, that can increase food production and supply.

Irrigation

Irrigation takes place when there is insufficient water or when water is not available during the growing season. Huge pivotal sprinklers – which tend to be wasteful – are used. Another method is drip irrigation (pipes on the ground deliver water direct to the roots of individual plants).

Irrigation commonly involves the local abstraction of water from underground aquifers or nearby rivers. It can also involve lengthy water transfers, using canals or pipelines such as the Indira Gandhi Canal in Rajasthan, India, and the Indus River Irrigation System, Pakistan (see page 150).

Aeroponics and hydroponics

Aeroponics and **hydroponics** use modern scientific techniques to grow crops without using soil. Closely controlled by scientists, plants can be grown throughout the year in artificially lit and heated greenhouses. The plants grow quickly as nutrients are applied directly to the roots, and diseases found in soils are eliminated.

However, these methods are expensive and require a great deal of expert knowledge.

The new green revolution use of biotechnology

The original 'green revolution' was hugely successful in introducing modern farming practices to parts of the developing world in the 1950s and 1960s. It revolutionised farming in countries such as India, averting famines.

Today's **'new' green revolution** involves a more sustainable and environmentally friendly approach; for example, promoting nutrient cycling through crop rotation, and mixed arable and livestock farming in less fertile areas.

In 2006, India promoted a second green revolution involving water harvesting techniques (collecting and storing water), soil conservation, the use of new seeds and livestock and rural transport developments to improve food distribution. A great deal of attention is now being focused on Africa, with its massive projected population growth.

Biotechnology

Biotechnology is a controversial scientific approach aimed at increasing yields by modifying products or processes. The development of genetically modified (GM) crops can increase crop production while using fewer chemicals and emitting less carbon dioxide. GM crops are grown in many parts of the world, for example maize in the Philippines and rapeseed in Canada. Over 50 per cent of the world's soya beans are GM.

While GM crops may not always increase food production directly, they can increase farmers' incomes, so they can purchase more food or invest in new food production methods on their own farms.

Appropriate technology

Appropriate (or intermediate) technology is a low-tech solution that makes use of local people and local cheap or recycled materials. This strategy is commonly used in LICs, often promoted by non-government organisations (NGOs) and charities. It may involve simple water harvesting techniques, irrigation schemes or crop processing, such as using bicycle power to de-husk coffee beans.

Now test yourself
TESTED

1 What is the difference between aeroponics and hydroponics?
2 How does the 'new' green revolution differ from the original green revolution?

Revision activity

Create a series of flashcards to summarise each of the strategies aimed at increasing food supply.

Example

Large-scale agricultural development

Learn about **one** example of a large-scale agricultural development, focusing particularly on its advantages and disadvantages.

Almeria, Spain

Almeria is an arid region in southern Spain. Since 1980, it has developed the largest concentrations of greenhouses in the world covering some 26,000 hectares. The greenhouses are used to grow a variety of fruit and vegetables, supplying countries such as the UK with out-of-season tomatoes, lettuces, cucumber and peppers. Over 50 per cent of Europe's fruit and vegetables are grown here, mostly using hydroponic technology.

The mild and sunny climate means that crops can be grown in unheated greenhouses, cutting production costs. Much of the labour involves relatively cheap foreign immigrants.

Advantages	Disadvantages
• Drip irrigation and hydroponic water recycling reduces water use in this dry environment. • Warm temperatures reduce energy costs. • Jobs are created in the greenhouses and in packing and transportation. • New scientific food-related companies have been attracted to set up in the area, boosting the economy and creating jobs. • Provides fresh fruit and vegetables throughout the year.	• Immigrant labour is low paid and living conditions are poor. • Social clashes between immigrants and local people. • Vast areas of land have been covered by plastic, affecting habitats and ecosystems. • Plastic waste is a major issue, some of which is dumped into the sea. • Use of pesticides is affecting human health. • Some natural water sources are under stress.

Indus Basin Irrigation Scheme, Pakistan

The Indus River is one of the world's great rivers, flowing through Pakistan on its way to the Arabian Sea. The Indus Basin Irrigation Scheme (IBIS) is one of the largest and oldest irrigation schemes in the world. First developed during British rule (1847–1947), it consists of three large dams and over a hundred smaller dams that regulate the water flowing through the river basin. Thousands of kilometres of canals and over 60,000 kilometres of ditches distribute the freshwater across the countryside to irrigate Pakistan's rich and fertile fields.

Advantages	Disadvantages
• Over 14 million hectares of land are under irrigation, adding 40% more land for cultivation. • Crop yields have increased enormously (e.g. wheat 36% and rice 38%), increasing Pakistan's food security. • Undernutrition (a lack of a balanced diet, i.e. lacking certain specific nutrients) in poorer regions has reduced due to diet improvements. • Large dams generate hydro-electric power (HEP). • Agricultural industries have been stimulated, boosting the economy and creating jobs.	• High temperatures cause water loss due to evaporation. • In some places, excessive irrigation has resulted in salinisation. • Downstream, people are deprived of water because others take too much. • Construction and maintenance costs are high – reservoirs silt up and canals need regular repair. • Rapid population increase will put pressure on the scheme in the future.

Exam practice

1 Describe how aeroponics and hydroponics can increase food supply. (4 marks)
2 Using a named example, evaluate the success of a large-scale agricultural development. (6 marks)

ONLINE

22.3 Sustainable food supplies

What is a sustainable food supply?

A **sustainable food supply** involves producing food without causing any damage to the natural environment. It involves and benefits local communities, supports the local economy and can be applied to HICs and LICs/NEEs. There are several strategies that can be adopted.

Organic farming

Organic farming involves the production of food without the use of chemicals, such as pesticides, insecticides and artificial fertilisers. It is widespread across the world, in both HICs and LICs/NEEs.

Organic food production is in harmony with nature and reduces the harmful effects of chemicals. However, it tends to be more labour intensive (e.g. weeding by hand) and yields are usually lower than more intensive forms of farming. This means that the produce can be more expensive to the consumer.

Permaculture

Permaculture (permanent agriculture) is similar to organic farming in that it promotes farming that is in harmony with the natural environment. It advocates using natural systems rather than artificial chemicals, for example using ladybirds as natural predators for aphid control (aphids can decimate green crops), rather than chemicals.

Permaculture promotes a sustainable lifestyle involving practices such as **sustainable development**, rainwater harvesting, composting, crop rotation and woodland management. Though widespread throughout HICs, permaculture has yet to be adopted widely in LICs/NEEs.

Urban farming initiatives

Urban farming often takes the form of community initiatives involving the conversion of waste or derelict land into productive farmland or vegetable gardens. Throughout the world, people living in urban areas grow food on rooftops, in patio gardens and in back yards.

The benefits of urban farming include increased food security, healthier diets, improved natural environments (with birds, butterflies and bees) and greater social cohesion.

People in Detroit, USA, have benefited from the Michigan Urban Farming Initiative, where derelict wasteland has been converted into garden beds, providing employment opportunities and fresh food

for local people. The Intercontinental New York Barclay Hotel has an apiary on its roof, with its bees producing honey that is used in the hotel's kitchen.

Fish and meat from sustainable sources

Increasingly, people are concerned about food sourcing; they want to know where their food comes from. There is considerable public demand, particularly in HICs, for food – especially meat and fish – to come from sustainable sources. Food labels now mostly provide this information, giving consumers the power to make a difference. For example:

- Sustainable fishing conserves fish stocks and avoids harmful over-fishing which can have catastrophic effects on aquatic ecosystems. The European Union applies strict fishing controls (quotas) to its members. Examples of sustainable fishing include the use of poles and lines rather than nets, diving for shellfish and returning fish and crustaceans that are below a certain size to the sea.
- Sustainable livestock rearing commonly involves pasture-fed, low-intensity systems with the minimal use of chemicals or vaccinations. Free-range systems, particularly involving chickens or pigs, are common throughout the world. These benefit the environment and provide higher standards of animal welfare. They are much more sustainable than the high-energy, intensive indoor systems operated in some parts of the world.

Seasonal food consumption

Due to refrigerated storage and air transport, people can buy food throughout the year regardless of the season. However, these high-energy systems are unsustainable and contribute to greenhouse gas emissions.

Increasingly, people are turning to local farmers' markets to buy seasonal produce, for example strawberries in the summer and apples/pears in the autumn. This reduces food miles and carbon footprints. Buying seasonal food also supports local economies and provides jobs.

Reduced waste and losses

Almost one-third of all the food produced is wasted, particularly in HICs. Over 60 per cent of food waste comprises fruit and vegetables, because these tend to go off quickly and require expensive storage and transportation systems.

Reducing food waste would have a huge positive impact on global food shortages. It can be achieved by:

- introducing refrigerated storage where none exists
- improving transport infrastructure
- processing food to lengthen its shelf–life (or preserving it by making jams and chutneys)
- applying common sense to consuming food beyond its 'best before' date.

The lack of adequate dry and chilled storage can lead to crops such as rice and wheat rotting, or becoming infested with vermin such as rats. This loss of food in areas already facing food shortages (mainly LICs) can be devastating. Investments in storage and transport can address this problem.

Now test yourself

TESTED

1 Define the following terms: sustainable food supply, organic farming, and permaculture.
2 List some urban farming initiatives.
3 How can meat and fish be obtained from sustainable sources?
4 What is the difference between food waste and food loss?

Revision activity

Create flashcards or produce a summary diagram to identify the main points about each of the strategies aimed at increasing sustainable food supplies.

Example

Local scheme in LIC/NEE to increase sustainable food supplies

You need to learn about **one** example of a local scheme in an LIC/NEE to increase sustainable supplies of food. You can choose one of the options below or use one other that you have studied.

Makueni Food and Water Security Programme, Kenya

Makueni is a rural country with a total population of just under 900,000 people, about 200 kilometres to the southeast of Nairobi, Kenya. Most people grow a variety of crops on the rich volcanic soils including millet, maize and sweet potatoes. However, the low and unreliable rainfall, averaging just 500 millimetres a year, causes crop failures, which in turn cause higher food prices and food shortages.

The Makueni Food and Water Security Programme aims to increase food and water security and increase sustainable supplies of food. In 2014, the African Sand Dam Foundation, together with the charity 'Just a Drop', introduced the programme in two villages, Musunguu and Muuo Wa Methovini.

Sand dams were constructed to filter and store rainwater in the ground. Together with a

rainwater harvesting tank, villagers now have a reliable source of water to irrigate their crops and provide drinking water for their livestock. Newly planted trees reduce soil erosion and increase biodiversity. The boost in crop yields has increased food security. The low-tech nature of the project ensures that it will increase sustainable supplies of food well into the future.

Jamalpur, Bangladesh

Jamalpur is an agricultural district in northern Bangladesh. Many farmers are subsistence farmers growing rice and wheat. The charity Practical Action has been supporting local farmers in introducing a new practice called rice-fish culture.

Small fish are introduced into the rice paddy fields, not only to provide a source of protein (for people eating the fish) but also to aerate the water and fertilise the soil (through their droppings). Rice yields have increased and family diets have improved. Surplus rice can be sold at local markets, providing extra income for farmers.

Rice-fish culture is a good example of a sustainable approach to increasing food supplies as it does not involve any artificial chemicals, expensive equipment or external expertise, and it does not harm the natural environment.

Exam practice

1 What is meant by a sustainable food supply? (2 marks)
2 Explain how urban farming initiatives and organic farming can increase sustainable food supplies. (6 marks)
3 Use an example of a local scheme in an LIC/NEE to explain how sustainable supplies of food can be increased. (6 marks)

ONLINE

23 Water

23.1 Water: supply and demand

What are global patterns of water surplus and deficit?

Figure 23.1 shows the global patterns of water surplus and water deficit. In surplus areas (e.g. much of North America, Europe and Asia) there may be high rainfall or low population densities. In areas of deficit (e.g. large parts of Africa, the Middle East and Australia) there may be low rainfall, or high population densities or agricultural demands.

You should be familiar with other terms associated with global water supply: **water security**, **water insecurity**, water quality and water stress. Areas of water stress include sub-Saharan Africa, parts of Asia including India, and large tracts of southern Europe.

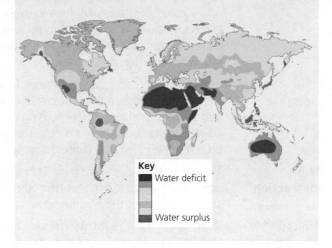

Key
Water deficit

Water surplus

Figure 23.1 Global patterns of water surplus and deficit

What are the reasons for increasing water consumption?

Water consumption has increased significantly in recent decades due to increasing demand from agriculture (for irrigation), industrial development and domestic use (especially in HICs).

Economic development

As economies develop the demand for water increases. The greatest demand for water is in HICs (e.g. the USA and Australia) whereas the lowest demand is in LICs. As the LICs and NEEs develop further, the demand for water will increase significantly:

- Agriculture – intensive farming requires huge quantities of water. With demand for food expected to increase by 70 per cent by 2050, increasing amounts of water will be required for irrigation and food processing. Agriculture is by far the largest consumer of water.
- Industry – processing and manufacturing industries (e.g. food processing, paper and textiles) use a huge amount of water. Water pollution can also be an issue, reducing the amount of water available for other purposes.
- Urbanisation – as the world becomes increasingly urbanised, the demand for water for drinking, washing and sanitation increases. Much of the

potential for future urbanisation is in areas already suffering from water stress, such as large parts of Africa and parts of Asia.
- Wealth – as people become richer their demands for water increase (e.g. domestic appliances, watering golf courses, and so on).

Rising population

Population growth increases the pressure on water supply. The world's population has grown rapidly since 1950 to its current level of about 7.5 billion. Almost all of this growth has been in LICs and NEEs – many of these countries suffer from a water deficit or water stress.

The fastest population growth is in Africa (2.51 per cent per year). This is the continent suffering most from water insecurity and the provision of safe water in the future will be a huge challenge.

Now test yourself

1 Describe the global pattern of water deficit.
2 Outline the reasons why economic growth and population growth increase water consumption.

TESTED

What factors affect water availability?

REVISED

Factor	Effects of water supply
Climate	Regions experiencing high rainfall totals tend to have a water surplus. However, water needs to be stored, either in underground aquifers or in lakes and reservoirs, before being transported. Some tropical LICs may receive adequate amounts of rainfall, but it may not be where it is needed or be high quality. Areas of low rainfall may experience water stress.
Geology	Aquifers are extremely important sources of water in many parts of the world, for example the Middle East (Jordan, Saudi Arabia and Egypt) and the UK, where water is stored within chalk beneath London and elsewhere within sandstone.
Pollution of supply	Pollution can lead to water stress. Causes of pollution include sewage and domestic waste, industrial effluent and agricultural runoff (containing chemicals or animal waste). Water pollution is a major problem in some LICs and NEEs where controls on pollution are limited.
Over-abstraction	This is a major problem with some of the world's aquifers formed during wetter climates in the past, such as parts of the Middle East.
Limited infrastructure	Many LICs lack the infrastructure to transport water. Pipelines and canals are expensive to construct and maintain; pumping stations require energy. Water transfer is also an issue in some NEEs and HICs, for example the USA, UK and China, where areas of water surplus are distant from the areas of greatest demand.
Poverty	Poverty often prevents people from access to safe water. Many poor people only have access to (potentially polluted) water from wells, or they have to pay for bottled water.

Now test yourself

Explain how climate and geology affect water availability.

TESTED

What are the impacts of water insecurity?

REVISED

Water insecurity can have serious social, economic and environmental impacts, particularly in LICs and NEEs.

Waterborne disease and water pollution

Many LICs and NEEs have inadequate sanitation systems:
- Rivers are often polluted from industrial and domestic waste.
- Open sewers can exist in the poorer parts of cities.
- Untreated drinking water can easily result in the spread of **waterborne diseases**.

This can result in serious illness and even death. Productivity is reduced and people's incomes fall.

The River Ganges in India is one of the world's most polluted rivers, containing raw sewage, industrial effluent, domestic waste and the ashes from cremated bodies (sacred Hindu funerals). Many people use the river for washing, irrigation and drinking water.

Food production

Agriculture is by far the major consumer of water:

- Countries experiencing water insecurity will be unable to irrigate crops and will suffer from low food productivity. This could have a serious impact on people's lives.
- Arid and semi-arid areas that experience low or unreliable rainfall are most at risk. The USA produces about a third of the world's staple food crops (wheat, maize and rice) yet much of its Mid-Western grain belt is prone to periodic drought. Food production dropped significantly in several years during the 1980s as a result of drought.

Industrial output

Several processing and manufacturing industries depend heavily on water:

- Examples include metal smelting, food processing, paper and textiles.
- Some forms of electricity production (e.g. nuclear) require water for cooling, or, in the case of hydro-electric power, for turning turbines.
- Lack of water can have serious economic impacts if, for example, electricity supply is unreliable.

Potential for conflict where demand exceeds supply

With increasing demand on water supplies, **water conflicts** often arise between neighbouring countries. This occurs when water sources – such as groundwater aquifers and rivers – are crossed by national borders. International politicians consider future 'water wars' to be a serious possibility.

Countries further down the course of a river may be adversely affected by reservoir construction or water abstraction higher up the river's course. The River Nile passes through seven countries before reaching Egypt, and the River Ganges passes through India before flowing into Bangladesh.

The Nurek Dam in Tajikistan has reduced water supplies on the Vakhsh River that flows into Uzbekistan. This has impacted on the cultivation of cotton in a country that already suffers from a water deficit. A new dam – the Rogun Dam – is likely to be constructed further upstream which could further reduce water flowing into Uzbekistan.

The River Jordan flows through Syria, Lebanon, Israel and Jordan, all countries experiencing water insecurity and simmering political conflict. They are all competing for vital water supplies.

Exam practice

1 Study Figure 23.1. Describe the patterns of water surplus and water deficit. (4 marks)
2 Explain how economic development causes an increase in water consumption. (6 marks)
3 Suggest the impacts of water insecurity on people's lives. (6 marks)

ONLINE

23.2 Water: increasing supply

What strategies can be used to increase water supply?

Water diversion and increasing storage

In some parts of the world, high levels of evaporation can seriously deplete water supplies. It is possible to divert surface water and pump it underground to be stored in aquifers where it will not be lost through evaporation. In Oklahoma, USA, rainfall is stored within the deep underlying alluvial soils so that it can be used when water supplies are low.

Dam and reservoir construction

Dams control water flow by creating reservoirs. Water can be stored during periods of water surplus and released when it is needed further downstream, say for irrigation. Dams can be used to generate electricity (hydro-electric power) and help prevent river flooding downstream.

Dams and reservoirs are expensive to construct and maintain and they can be extremely controversial. Often people have to move home and valuable farmland may be flooded. The Three Gorges Dam in China flooded several settlements and displaced thousands of people. The largest reservoir in the UK is Kielder Reservoir in Northumberland, which is over 10 kilometres long. It regulates water flowing down the River Tyne and also generates electricity.

Water transfers

Water transfer schemes – involving pipelines or canals – can move water from areas of surplus to areas of deficit. Schemes can be expensive, but they are very effective and are common in both HICs and LICs/NEEs.
- In the UK, water is transferred by pipeline from wetter areas in Wales to large populations in the Midlands. Water from the River Severn is moved by canal to supply Bristol.
- In Australia, water is transferred between the Murray/Darling River basins.

See the examples on page 157.

Desalination

Desalination involves extracting salt from seawater to create fresh drinking water. It is an extremely expensive process, which explains why it is only really an option in HICs experiencing severe water insecurity, such as Saudi Arabia, Australia, the USA and Spain. In the future, desalination may become an option for some NEEs as the ever-increasing demand for water results in more expensive solutions.

Desalination can, however, have harmful environmental impacts. Salt waste can damage marine ecosystems and the high energy demand increases carbon emissions.

Now test yourself

1 How can dams and reservoirs increase water supply?
2 Why have some countries turned to desalination as a strategy to increase water supply?
3 Outline some of the problems associated with desalination.

TESTED

Example

Large-scale water transfer scheme

You need to learn about **one** example of a large-scale water transfer scheme, focusing particularly on its advantages and disadvantages.

South–North Water Transfer Project, China

The South–North Water Transfer Project (SNWTP) is one of the world's most expensive and ambitious water transfer projects. The aim of the project is to move water from the relatively wet south of the country to the more arid north.

The water is needed to support economic development (agriculture and industry) on the North China Plain and the rapid growth of Beijing and Tianjin. Increased demand from the 200 million people living in the region has depleted groundwater aquifers, causing the water table to fall by up to 5 metres each year.

The US$62 billion scheme involves moving water along three distinct routes from the Yangtze River basin in the south to the Yellow River basin in the north.

Advantages	Disadvantages
• The scheme reduces water insecurity in northern China and promotes economic development. • Water is made available for irrigation, increasing food security in the region. • Improvements in water quality will enable people to enjoy better health and be more productive at work. • Improved water supply will support industrial (and economic) growth.	• Large numbers of people displaced to enable reservoirs to be constructed (over 300,000 were displaced by the construction of the reservoir at Danjiangkou). • Ecological damage and environmental disturbance associated with the developments. • The region is prone to earthquakes that could cause expensive damage. • Very expensive project with taxpayers having to pay.

Lesotho Highland Water Project, Lesotho

Lesotho is a small highland country in southern Africa entirely surrounded by South Africa. While it is a poor country with low levels of development and some degree of food insecurity, it experiences heavy rainfall and enjoys a water surplus. Neighbouring South Africa, in contrast, has a large demand for water and suffers from a water deficit. Water is needed to support agricultural and industrial development.

The Lesotho Highland Water Project involves transferring water from the Segu River in Lesotho to the River Vaal in South Africa. The project will take about 30 years to complete and will involve the construction of several huge dams and reservoirs, pipelines and tunnels.

Exam practice

1 Using a named example, discuss the advantages and disadvantages of a large-scale water transfer scheme. (6 marks)

ONLINE

Advantages	Disadvantages
South Africa • A more reliable source of water for irrigation and industrial development. • The provision of safe water for those currently without access. • Ecological benefits to the river system due to the influx of large quantities of freshwater. **Lesotho** • Income from the scheme will account for 75% of gross domestic product (GDP). • Supplies the country with electricity (hydro-electric power), improving the quality of people's lives. • Improved road infrastructure. • Improvements to safe water provision and sanitation.	**South Africa** • The cost of the scheme is estimated at US$4 billion. • Leakages could account for 40% of the water. • Some poor people may not be able to afford the high water tariff costs. • Corruption has been an issue. **Lesotho** • Over 30,000 people have been displaced by dams and reservoirs, and some valuable agricultural land has also been lost. • Ecological destruction to make way for reservoirs. • Corruption has prevented compensation payments reaching people.

23.3 Sustainable water supplies

How are sustainable water supplies achieved?

A **sustainable water supply** does not cause any damage to the natural environment. It involves and benefits local communities, supports the local economy and can be applied to both HICs and LICs/NEEs. There are several strategies that can be adopted.

Water conservation

Water conservation involves reducing waste and using water more sparingly.

- Reducing leakages – up to a third of global water supply is lost through leaks, mainly due to broken pipes. Reduction in leakages would have a hugely beneficial effect on millions of people. In the UK, an estimated 3.3 billion litres of treated water is lost each year, enough to meet the daily needs of over 20 million people!
- Installations of water meters in homes and businesses to reduce water usage.
- Using water more efficiently in the home, for example by taking showers rather than baths and using appliances (washing machines, dishwashers) that do not use excessive amounts of water.
- Using more efficient irrigation techniques, such as drip irrigation (direct to each plant) rather than wasteful sprinkler systems, where a high proportion of water is evaporated.
- Preventing water pollution of rivers and aquifers (caused by industrial effluent and agricultural chemicals), thereby conserving freshwater supplies.

> **Revision activity**
>
> Consider ways in which your family or local community conserve water.

Groundwater management

Groundwater management is essential if aquifers are to be sustainable, both in terms of water quantity and water quality. Water abstraction needs to be balanced either by natural recharge (precipitation) or by artificial recharge, pumping water underground from rivers and lakes. There are huge problems in some parts of the world, for example in North Africa and the Middle East, where 'fossil' aquifers formed during past wetter climates are now being exhausted.

In HICs, groundwater management by water companies or local authorities is usually quite effective, involving monitoring of water tables and water quality. Regulations are imposed on water abstraction to ensure that aquifers remain healthy and that surface water sources (particularly rivers) are not adversely affected. Ineffective management can result in rivers drying up as water tables fall, having serious impacts on ecosystems.

In LICs, many families and communities rely on unregulated groundwater accessed from the surface by wells. Pollution is common, due to the lack of sanitation, and these water sources can become severely depleted if overused.

Water recycling

It is said that drinking water in London has been through as many as eight people before reaching you! The treatment and re-use of water is common throughout the world. Domestic or industrial wastewater can be treated and used for a range of purposes including irrigation, electricity generation (cooling) and industrial purposes.

'Grey water'

Domestic waste water ('**grey water**') is increasingly being recycled and used both inside and outside the home. Logically, it makes no sense to use expensively treated drinking water to flush the toilet or water the garden!

In water deficit areas (e.g. Spain), modern houses are being constructed with separate drinking water and 'grey water' systems to help to conserve and recycle water efficiently. Such systems are expensive to install, so this tends to happen in HICs only.

Now test yourself
TESTED

1 What is meant by a 'sustainable water supply'?
2 Why do groundwater supplies need to be well managed?
3 What is 'grey water' and how can its use increase water supplies?

Example

Local scheme in LIC/NEE to increase sustainable supplies of water

You need to learn about **one** example of a local scheme in an LIC/NEE to increase sustainable supplies of water. You can choose one of the options below or use one other that you have studied.

Hitosa, Ethiopia

Hitosa is a semi-arid rural area about 160 kilometres south of Addis Ababa, the capital of Ethiopia. In the 1990s a 140-kilometre gravity-fed water transfer scheme was installed, moving water by pipeline from mountain springs on the slopes of Mount Bada to Hitosa. The safe water is made available to households using public standpipes and it is also used by farmers to irrigate fields of wheat and barley. The project is managed by the local communities.

The scheme – partly funded by the charity WaterAid – has been extremely successful, providing a reliable and safe water supply for over 65,000 people. Time spent collecting water has been reduced and new businesses (such as cattle fattening) have become established.

Wakel River Basin Project, Rajasthan, India

The Wakel River Basin is in southern Rajasthan, a poor desert region in northern India. The semi-arid area suffers from water insecurity as over-abstraction for irrigation has reduced groundwater supplies.

Supported by the US Agency for International Development, the project aims to increase water supply by improving storage and ensuring better management practices. One aspect of the project at the local community level encourages more effective methods of rainwater harvesting:

- Underground storage tanks (3 metres in diameter and about 3 metres deep), called 'taankas', to store water and prevent it evaporating.
- Small earth dams, called 'johed', to interrupt the surface flow of the water and encourage it to soak into the soil, thereby raising the water table.

To improve water management, farmers use small dams to block river channels, diverting water into fields for irrigation. This is managed by local communities to ensure that water is distributed evenly and fairly. Maintained by farmers, it is entirely sustainable.

Exam practice

1 What is meant by a sustainable water supply? (2 marks)
2 Explain how water conservation and groundwater management can increase sustainable supplies of water. (6 marks)
3 Use an example of a local scheme in an LIC/NEE to explain how sustainable water supplies can be increased. (6 marks)

ONLINE

24 Energy

24.1 Energy: supply and demand

What are the global patterns of energy? REVISED

Energy is required for a wide range of activities including heating, lighting, producing food, powering industry and fuelling transport. As a country develops, ever-increasing amounts of energy are required.

Figure 24.1 shows the global patterns of energy consumption and energy supply:

- Countries with rich energy resources (e.g. countries in North America, Europe, Australasia and parts of the Middle East) also have high energy consumption. Mostly these are HICs.
- Some regions (e.g. South America, North Africa and parts of Asia) have high energy consumption but limited energy supply. Most of these countries have to rely on imported energy.
- Much of central and western Africa experiences low energy consumption and these regions also have scarce resources.

It is the relationship between energy supply and demand (consumption) that largely determines **energy security**. Notice that most of the world experiences medium or high levels of insecurity (see Figure 24.2):

- High levels of energy insecurity are experienced by both HICs (e.g. the USA, UK and many other European countries) and LICs/NEEs (e.g. many countries in Africa, the Middle East, South America and Asia).
- Countries with the lowest levels of insecurity tend to be those richly endowed with resources, where supply exceeds demand (e.g. Canada, Russia and Saudi Arabia).
- Energy insecurity can be experienced by countries with both a high and low energy consumption.
- Even some countries with a high energy supply (e.g. the USA) can experience energy insecurity where demand for energy is high.

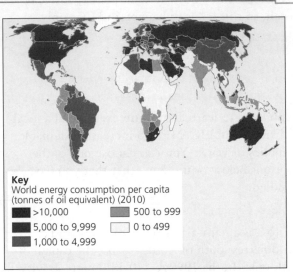

Key
World energy consumption per capita (tonnes of oil equivalent) (2010)
- >10,000
- 5,000 to 9,999
- 1,000 to 4,999
- 500 to 999
- 0 to 499

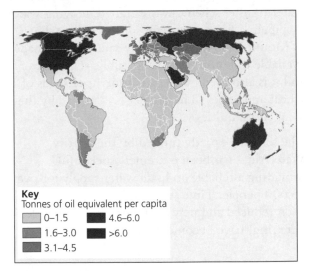

Key
Tonnes of oil equivalent per capita
- 0–1.5
- 1.6–3.0
- 3.1–4.5
- 4.6–6.0
- >6.0

Figure 24.1 Global patterns of energy consumption and energy supply

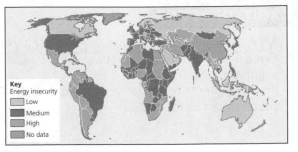

Key
Energy insecurity
- Low
- Medium
- High
- No data

Figure 24.2 Global pattern of energy insecurity

Now test yourself TESTED

1 Describe the global pattern of energy consumption.
2 What is meant by the term energy security?

Why is energy consumption increasing?

Energy consumption has increased significantly in recent decades due to rising demand for it in industrial development, transport, agriculture and domestic use (especially in HICs).

Economic development

As economies grow and develop, the demand for energy increases. The greatest demand for energy is in HICs (e.g. the USA and Australia) and rapidly developing NEEs (e.g. China, Brazil, Nigeria and Malaysia). For the future, energy demand is likely to increase in LICs, particularly in Africa.

- Agriculture – intensive farming requires huge quantities of energy to power machinery and provide lighting and heating. The manufacture and transport of agricultural foodstuffs, chemicals and fertilisers also use energy. As farming becomes more intensive across many parts of the world, energy use will increase.
- Industry – processing and manufacturing industries use a huge amount of energy, mostly in the form of electricity. Rapid industrialisation is taking place in NEEs and may begin to take place in LICs in the future, increasing global demands for energy.
- Transport – despite the recent development of more efficient engines, energy use in vehicle production is high and likely to increase as improvements in living standards result in greater car ownership.
- Urbanisation – as the world becomes increasingly urbanised, the demand for energy for heating, lighting and for powering domestic appliances will increase. Much of the potential for future urbanisation is in areas already suffering from energy insecurity, such as large parts of Africa and parts of Asia.
- Wealth – as people become richer, the demand for energy increases (e.g. domestic appliances, computers, leisure and recreational activities, etc.). The future development of populous NEEs such as China, India and Brazil will significantly increase demand on global energy supplies.

Rising population

Population growth increases the pressure on energy supplies. The world's population has grown rapidly since 1950 to its current level of about 7.5 billion. Almost all of this growth has been in LICs and NEEs – many of these countries suffer from an energy deficit.

The fastest population growth is in Africa (2.51 per cent per year). This is the continent suffering most from energy insecurity. The provision of energy in the future will be a huge challenge.

Technology

Improvements in technology have both positive and negative impacts on energy supply and consumption:

- Technology enables **fossil fuels** to be extracted from challenging environments, such as from deep below oceans or from shale rock (fracking).
- Renewable energy technology has developed hugely in recent years to become much more efficient and cost-effective (e.g. wind turbines, solar farms).
- Improvements in the efficiency of vehicle engines have reduced fuel consumption and harmful emissions, although car ownership is growing rapidly.
- **Energy conservation** is much more effective (e.g. heating and lighting) due to technological advances, both in the home and the workplace.
- Technology has created a greater range of products that use energy, particularly in the home.

Now test yourself

TESTED

Outline the reasons why economic growth and population growth increase energy consumption.

Revision activity

Draw up a summary table with three columns to identify how technology can lead to:
a an increase in energy supply
b an increase in energy demand
c a decrease in energy demand.

What factors affect energy availability?

Factor	Effects on energy supply
Physical factors	• Geology determines the availability of energy sources (coal, oil and natural gas). • In some areas (e.g. Siberia and Alaska), climate and relief can create massive challenges for energy supply and transportation. • Substantial reserves of oil and natural gas lie beneath deep and dangerous oceans. • Physical factors provide some opportunities for renewable energy, such as plentiful water and mountains for **hydro-electric power**.
Cost of exploitation and production	• The cost of exploitation and production often determines whether energy reserves can be exploited. These costs fluctuate depending on demand and supply. • Where costs are low, energy is cheap, and so demand can grow. The opposite is also true.
Technology	• Developments have enabled new reserves of energy to be exploited (e.g. the increased use of fracking). • New technologies have led to the growth in renewables, with developments in solar, wind and geothermal sources.
Political factors	• Energy insecurity has resulted in global conflicts (for example, in the Middle East). • Conflicts can interrupt energy flows, e.g. recent civil war in Libya has reduced oil exports. • In order to secure energy sources, high-consumer HICs often promote friendly or supportive relationships with foreign countries.

What are the impacts of energy insecurity?

Energy insecurity can have serious social, economic and environmental impacts.

Exploration in sensitive environments

As energy demand increases, exploration threatens to harm sensitive natural environments:

- Drilling for oil and gas in Alaska and Siberia threatens highly fragile tundra and polar environments.
- In tropical regions, rainforests have been cleared to make way for biofuel production.
- Valleys have been flooded to create reservoirs to generate hydro-electric power (e.g. Three Gorges Project in China), displacing thousands of people and flooding valuable land.
- Construction of wind and solar farms are very controversial, especially in attractive landscapes such as the English Lake District or Scottish Highlands.

Economic and environmental costs

The exploitation of energy from marginal and environmentally sensitive regions is expensive and will increase the cost of energy. This will lead to rising prices for consumers.

Food production

Food production increasingly depends on high energy inputs, such as animal feed, powering machinery, and manufacturing (and transporting) fertilisers to maximise production. Energy insecurity and rising prices will affect food production and increase consumer costs.

> **Revision activity**
>
> Draw a spider revision diagram to summarise the main factors affecting energy supply.

> **Now test yourself**
>
> Explain how physical geography can affect energy supply.
>
> TESTED

Industrial output

Energy insecurity and fluctuating prices can have devastating impacts on industrial production, leading to uncertainty and higher prices. Competition with other countries may lead to industrial closures and job losses. Power cuts are common in many countries suffering from energy insecurity, including newly emerging economies (NEEs) such as Nigeria. This reduces industrial output.

Conflict

- In countries with energy insecurity, the different consumer sectors, such as agriculture, industry and domestic users, may be in conflict. Governments may have to decide on energy use priorities.
- A market may become flooded by cheap products from countries with high levels of energy security. This might lead to trade tariffs and other controls to protect a country's home industries.
- Several recent conflicts between countries have been linked to energy insecurity, for example political disputes along the route of the pipelines transporting gas from Russia to the West, through countries such as Ukraine.

Exam practice

1 Study Figure 24.2. Describe the global pattern of energy insecurity. (4 marks)
2 Explain how economic development causes an increase in energy consumption. (6 marks)
3 Suggest the economic and environmental impacts of energy insecurity. (6 marks)

ONLINE

Revision activity

Draw up a table to summarise the social, economic and environmental impacts of energy insecurity.

Exam tip

With Question 3, make sure you focus clearly on economic and environmental impacts and ensure that your answer is reasonably well balanced.

24.2 Energy: increasing supply

What strategies can be used to increase energy supply? REVISED

There are two main strategies for increasing energy supply:

- Continue to develop non-renewable fossil fuels and **nuclear power**. This could involve exploiting new reserves in increasingly challenging environments or using technology to improve methods of fossil fuel extraction, such as fracking.
- Develop further alternative renewable sources of energy, which are less damaging to the environment and more sustainable in the long term.

Energy is mostly delivered to the consumer in the form of electricity. This can be generated by burning fossil fuels (see Figure 24.3) in thermal power stations or mechanically, for example using wind or water to turn turbines.

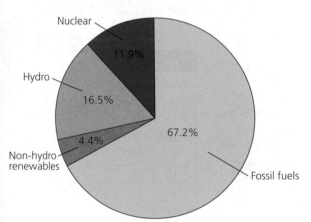

Figure 24.3 Fuels used to generate electricity

Non-renewable

Some fossil fuels can be mined near to the ground surface. However, mining most commonly uses complex and expensive technologies to extract the energy sources from deep underground.

Eventually, non-renewable energy sources will become too expensive (economically or environmentally) to be extracted. In the meantime, there are still significant amounts available. Despite their emissions of harmful carbon dioxide, it seems likely that fossil fuels will remain important for some time to come.

Nuclear power accounts for about 12 per cent of global electricity generation. It is a non-renewable energy source as it uses uranium (a rare mineral) as a raw material. Nuclear fission heats water to create steam which drives the turbines to generate electricity.

France relies heavily on nuclear power. The UK has recently commissioned a new power plant at Hinkley Point, Somerset. Despite its efficiency in producing long-term energy, nuclear power stations are very expensive to construct and operate. There are also environmental concerns about nuclear waste, which can remain harmful for centuries.

Renewable

Renewable energy sources currently account for about 21 per cent of global electricity generation. Over 75 per cent of this comes from hydro-electric power (HEP). See below for some advantages and disadvantages of renewable energy sources:

Renewable energy source	Advantages	Disadvantages
Biomass	• Can involve low-tech organic matter (wood, manure). • Can use vegetation grown in different climatic zones.	• Controversial in that tropical rainforests are cut down to make way for commercial biofuels. • Burning organic matter contributes to carbon emissions.
Wind	• Wind farms have considerable potential – in the UK wind energy accounts for 10% of electricity. • Manufacturing and maintenance creates jobs and boosts the economy.	• Turbines are expensive to manufacture and maintain. • Some consider them to be unsightly – there are often strong objections to land-based wind farms.
Hydro-electric power (HEP)	• Very efficient and effective. • Well suited to remote mountainous environments. • Micro-hydro schemes can supply small isolated communities. • Reservoirs can supply water for irrigation and help control flooding.	• Large dams and reservoirs are expensive and controversial. • Local people may be displaced and valuable land flooded.

Renewable energy source	Advantages	Disadvantages
Tidal	• Could be effective in areas experiencing a high tidal range, e.g. Severn Estuary, UK. • Tidal barrages may help protect coastlines from sea level rise or storm surges.	• Very expensive. • May affect marine ecosystems and fish migration.
Geothermal	• Extremely effective in volcanic areas, e.g. Iceland, which produces huge quantities of cheap energy. • Hot water can be used for industrial processes, swimming pools, fish farming, and so on.	• Some ecological harm may result from waste hot water. • Involves reservoir construction (to provide the freshwater used in electricity generation).
Wave	• Potentially useful in supplying local energy to coastal locations.	• Very expensive to construct. • Wave energy varies enormously from day to day.
Solar	• Areas with high sunshine totals (e.g. USA and Spain) make increasing use of solar farms. • Potentially very effective in converting sunlight into electricity.	• Dependent on sunshine amounts. • Expensive to construct. • Controversial use of fields instead of growing food.

Now test yourself

TESTED

1 What are fossil fuels?
2 How can nuclear power increase energy supply?
3 Describe how **two** examples of renewable energy can increase energy supply.

Example

Fossil fuel extraction – advantages and disadvantages

You need to study one example to show how the extraction of a fossil fuel has both advantages and disadvantages. The example below considers natural gas.

Natural gas extraction

Along with coal and oil, natural gas is a fossil fuel. It was formed by the decomposition of living organisms deposited on the seabed and buried millions of years ago. Trapped deep below the ground within shale deposits, it is extracted and transported primarily to power industry and to generate electricity.

Natural gas production is dominated by Russia, Iran, Qatar and Turkmenistan. Major gas pipelines transport the gas across Europe to markets in the West. Recently, shale gas has been exploited using the controversial fracking process. This involves rocks being fractured and injected with water and chemicals to extract the gas.

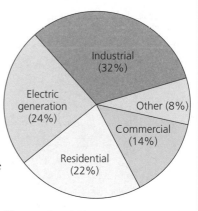

Figure 24.4 Natural gas use by sector

Advantages	Disadvantages
● Natural gas is significantly 'cleaner' than coal or oil, with carbon emission 45% less than coal and 30% less than oil. ● It does not create waste (such as coal ash). ● It disperses quickly in the air if there is a leak, whereas oil leaks can be hugely damaging to the environment. ● Produces electricity cheaply. ● Easily distributed by pipeline, direct to users. ● Versatile in that it can be used for heating and cooking. ● Huge global reserves.	● Leakages can result in explosions or fires; gas is toxic and harmful to human health. ● Greenhouse gases (especially carbon dioxide) are released, contributing to climate change and global warming. ● Fracking is controversial and has been linked to causing small earthquakes, ground subsidence and water pollution. ● It is naturally odourless, so can be undetected unless an odorant is artificially added. ● Infrastructure (pipelines) are expensive to construct and can be environmentally damaging. ● Political issues can disrupt transportation, particularly from Russia to the West.

Exam practice

1 Study Figure 24.3. Describe the pattern of fuels used to generate electricity. **(2 marks)**
2 Describe how fossil fuels can increase energy supply. **(4 marks)**
3 Define biomass and explain why it is a renewable form of energy. **(4 marks)**
4 Using a named example of a fossil fuel, suggest how its extraction has both advantages and disadvantages. **(6 marks)**

ONLINE

Revision activity

Draw a summary diagram to outline the advantages and disadvantages of the extraction of a fossil fuel.

24.3 Sustainable energy supplies

How can we move towards sustainable energy supplies?

A **sustainable energy supply** aims to balance supply and demand by:
- using fossil fuels efficiently
- developing renewable supplies
- reducing waste by conserving energy
- greater fuel efficiency in the home and in the workplace.

Through reducing the use of fossil fuels, and increasing levels of fuel efficiency and conservation, it is possible to reduce our carbon footprint.

A sustainable energy supply does not cause any damage to the natural environment. It involves and benefits local communities, supports the local economy and can be applied to both HICs and LICs/NEEs. There are several energy conservation strategies that can be adopted.

Now test yourself

TESTED

What is meant by a sustainable energy supply?

What are conservation strategies for a sustainable future?

REVISED

Designing homes and workplaces

Figure 24.5 describes ways that homes can be designed to conserve energy.

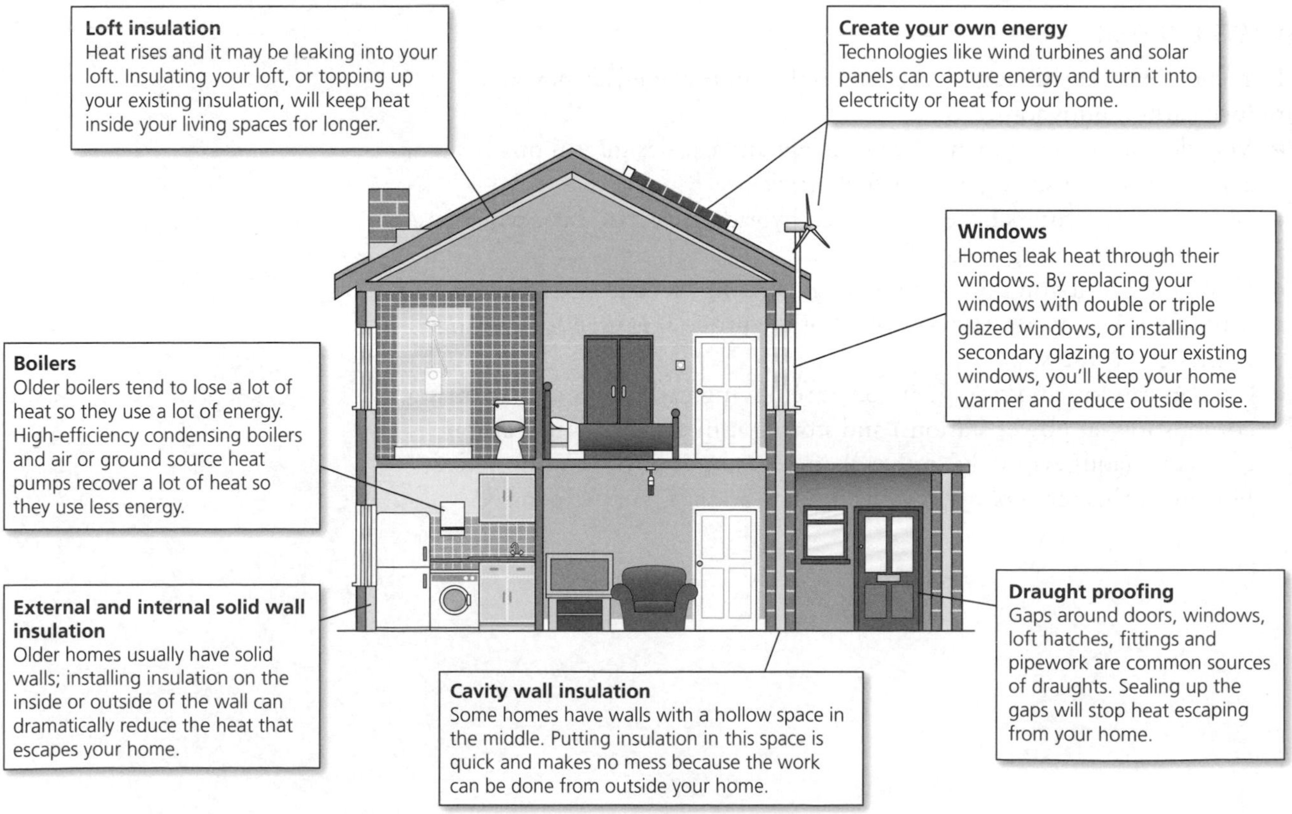

Loft insulation
Heat rises and it may be leaking into your loft. Insulating your loft, or topping up your existing insulation, will keep heat inside your living spaces for longer.

Create your own energy
Technologies like wind turbines and solar panels can capture energy and turn it into electricity or heat for your home.

Windows
Homes leak heat through their windows. By replacing your windows with double or triple glazed windows, or installing secondary glazing to your existing windows, you'll keep your home warmer and reduce outside noise.

Boilers
Older boilers tend to lose a lot of heat so they use a lot of energy. High-efficiency condensing boilers and air or ground source heat pumps recover a lot of heat so they use less energy.

External and internal solid wall insulation
Older homes usually have solid walls; installing insulation on the inside or outside of the wall can dramatically reduce the heat that escapes your home.

Cavity wall insulation
Some homes have walls with a hollow space in the middle. Putting insulation in this space is quick and makes no mess because the work can be done from outside your home.

Draught proofing
Gaps around doors, windows, loft hatches, fittings and pipework are common sources of draughts. Sealing up the gaps will stop heat escaping from your home.

Figure 24.5 Energy conservation in the home

Many of these initiatives can be applied to the workplace but at a larger scale. These can include:

- regulating thermostats
- ensuring doors and windows are closed when the heating is on
- turning off computers and other electrical appliances when not needed
- using low-energy lighting.

Transport

Transportation uses a great deal of energy through oil-based fuel (i.e. petrol and diesel). Strategies to use energy more efficiently include:

- using public transport rather than private cars (several authorities use buses powered by biomass and other alternative energy sources)
- encouraging people to buy hybrid or electric cars
- encouraging people to car-share – some cities have dedicated car-sharing lanes as an incentive
- reducing the use of air travel, especially for short-haul flights.

Reducing energy demand

In addition to the energy-saving measures described above – which will all lead to a reduction in energy demand – people can:

- be incentivised through government grants or tax relief to conserve energy in the home, e.g. by installing loft insulation or double-glazing
- be given financial incentives to encourage them to drive more efficient vehicles or to switch to hybrid or electric cars
- be encouraged to alter their behaviour by reducing the temperature of heating thermostats or domestic hot water
- use legislation to demand high levels of energy efficiency for new-build houses.

Using technology to improve efficiency of using fossil fuels

In recent years, technology has been used to improve efficiency and reduce carbon emissions:

- Vehicle engines are much more efficient and car design is more aerodynamic, ensuring greater fuel efficiency.
- Many manufacturers have introduced hybrid or electric cars, which are becoming increasingly popular.
- Combined heat and power involves generating electricity, rather than using fuel. Hot water (a by-product of the process) is used for heating homes or businesses.
- Carbon capture and storage involves removing carbon produced during combustion (at power stations) and storing it in underground carbon reservoirs (aquifers, rocks) rather than releasing it into the atmosphere. However, this technology is in its infancy and it is very expensive.

> **Revision activity**
>
> Make a list of five energy conservation strategies in the home. Include strategies that your family have adopted, as you are more likely to remember these in the exam!

Now test yourself and Exam practice answers at **www.hoddereducation.co.uk/myrevisionnotes**

Local renewable energy scheme in an LIC/NEE

You need to learn **one** example of a local renewable energy scheme in an LIC or NEE that provides a sustainable energy supply. You could choose one of the two examples given below or use another one that you have studied at school.

The two examples that follow consider micro-hydro schemes. These schemes are widely used in LICs/NEEs to provide a sustainable energy supply, often in remote rural areas, and to promote **sustainable development**. They are relatively cheap to install and can be maintained by local communities. They are adaptable to a wide range of locations and do not involve major environmental impacts. Micro-hydro schemes are renewable (using the power of flowing water) and do not emit carbon.

Ruma Khola micro-hydro, Darbang, Nepal

Over 1,000 micro-hydro schemes have been constructed in Nepal, supported by the government and the World Bank. Darbang is a remote settlement in the foothills of the Himalayas some 225 kilometres northwest of Kathmandu. A 51-kilowatt micro-hydro plant has been constructed primarily to supply electricity to 700 households in five villages, including Darbang. Since it became operational in 2009, it has led to an influx of small industries eager to make use of the electricity. These include furniture workshops, a cement block manufacturer and a noodle factory. Local farms have also benefited.

Chambamontera micro-hydro, Peru

Chambamontera is a small Andean mountain community in the north of Peru. It is a very remote and isolated community, surviving largely on subsistence farming. Until recently it had no electricity to supply power, light and heat.

The US$51,000 Chambamontera micro-hydro scheme was funded by the Peruvian government, Japan and the charity Practical Action. Local people also contributed towards the scheme. The steep slopes and high rainfall made it an ideal option as a long-term sustainable energy supply.

The scheme has brought many benefits to the local community:
- Reliable electricity provides street lighting, heating and power for domestic appliances such as fridges.
- People can use electricity for cooking rather than kerosene, which had harmful impacts on people's health.
- Children have electricity in school (for light and powering computers) and at home in the evenings for completing homework.
- Local industries have benefited from the power.
- Less fuelwood is required, resulting in less deforestation and soil erosion.

Revision activity

Draw a summary diagram to outline the main features of a local renewable energy scheme.

Exam practice

1 Describe how energy conservation in the home can contribute towards a sustainable resource future. (4 marks)
2 Explain how technology can increase the efficiency of using fossil fuels. (6 marks)
3 Use a named example of a local renewable energy scheme in an LIC/NEE to assess its success in providing a sustainable energy supply. (6 marks)

ONLINE

Unit 3: Geographical Applications

Assessment outline

The Unit 3 Geographical Applications exam paper assesses:
- issue evaluation
- fieldwork
- geographical skills.

The examination is 1 hour 15 minutes and is worth 76 marks. It is split into two broad sections:
- Section A: Issue evaluation (37 marks)
- Section B: Fieldwork (39 marks)

Geographical skills are assessed throughout this paper and also in Papers 1 and 2.

How to succeed in Unit 3 Geographical Applications

Section A: Issue evaluation

This section aims to test your critical thinking and problem-solving skills. It provides opportunities for you to demonstrate geographical skills and applied knowledge and understanding by looking at a particular geographical issue(s) using secondary sources.
- The issue(s) will arise from any aspect of the **compulsory** sections of the subject content but may extend beyond it through the use of resources in relation to specific unseen contexts.
- The knowledge and understanding that you have gained from studying Units 1 and 2 will support you in this synoptic assessment.

Resource booklet

A six-page resource booklet will be made available to your teacher **twelve weeks** before the date of the exam. This will enable you to become familiar with the material before the exam. You will not be allowed to take the original resource booklet into the examination room but will be issued with a clean copy in the exam.

The resource booklet will contain a variety of geographical information linked to a particular theme. Information could include maps at different scales, diagrams, graphs, statistics, photographs, satellite images, sketches, extracts from published materials, and quotes from different interest groups.

You should remember to refer to specific information in the resources, making use of information in the key, map evidence, scales and axes on graphs.

Be prepared to describe patterns, suggest trends and identify relationships.

Figure 25.1 You will need to interpret and analyse the information in the resource booklet

Exam assessment

The exam will consist of a series of questions related to a contemporary geographical issue(s), leading to a more extended piece of writing which will involve an evaluative judgement.

- Early questions – these will require you to interpret and analyse the information provided in the resource booklet.
- Final decision-making question – this final question in Section A will require you to apply your knowledge and understanding of the information in the resource booklet to develop a critical perspective on the issue(s) studied. You will need to consider the points of view of the stakeholders involved, make an appraisal of the advantages and disadvantages, and evaluate the alternatives. You may, for example, be asked to suggest an appropriate project for urban improvement or to identify a preferred new road or rail route in the UK. You will be expected to justify your decision using the evidence available.

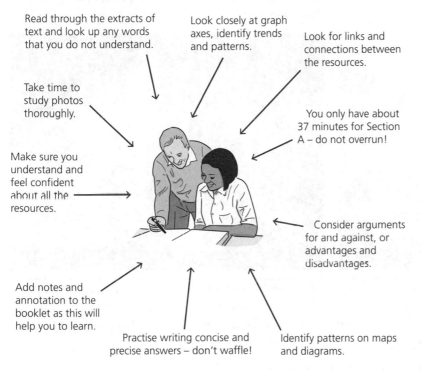

Read through the extracts of text and look up any words that you do not understand.

Look closely at graph axes, identify trends and patterns.

Look for links and connections between the resources.

Take time to study photos thoroughly.

You only have about 37 minutes for Section A – do not overrun!

Make sure you understand and feel confident about all the resources.

Consider arguments for and against, or advantages and disadvantages.

Add notes and annotation to the booklet as this will help you to learn.

Practise writing concise and precise answers – don't waffle!

Identify patterns on maps and diagrams.

Figure 25.2 Preparing for Section A

Section B: Fieldwork

As part of your GCSE course you will have completed two geographical enquiries, one physical and one human. Remember that in at least one of the enquiries you must show an understanding of both physical and human geography and their interactions. Make sure that you know which of your investigations involves this!

You will be expected to:

- apply knowledge and understanding to interpret, analyse and evaluate information and issues related to geographical enquiry
- select, adapt and use a variety of skills and techniques to investigate questions and issues and communicate findings in relation to geographical enquiry.

Exam assessment

In common with Section A, it is possible to identify two distinct sets of questions about fieldwork in Section B.

● Generic fieldwork questions – these are based on the use of fieldwork materials from an unfamiliar context (such as graphs and diagrams) for you to criticise, or, for example, information about a location or sample strategy for you to evaluate.

● Individual fieldwork enquiry questions – these questions are based on **your** two enquiries. You must learn the titles as you will be asked to write them down. The questions will focus on reasons and justification, so you must understand **why** you did things and not just what you did (Figure 25.3).

Explain how the theory behind the investigation determined the data collection method(s) used.

Justify the sampling strategies used in your enquiry.

Justify the choice of location(s) used to collect data.

Assess the appropriateness of your data collection methods.

Evaluate the effectiveness of your data collection methods.

Assess the appropriateness of your data presentation methods.

To what extent can the fieldwork results be deemed to be reliable?

Evaluate the accuracy and reliability of your results/conclusions.

Figure 25.3 Typical questions about individual fieldwork enquiries

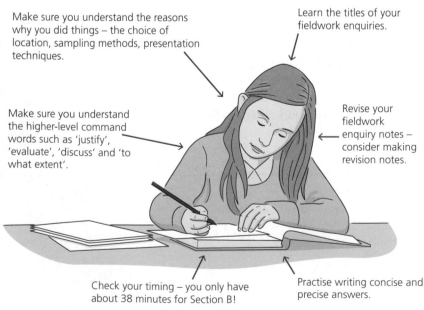

Make sure you understand the reasons why you did things – the choice of location, sampling methods, presentation techniques.

Learn the titles of your fieldwork enquiries.

Make sure you understand the higher-level command words such as 'justify', 'evaluate', 'discuss' and 'to what extent'.

Revise your fieldwork enquiry notes – consider making revision notes.

Check your timing – you only have about 38 minutes for Section B!

Practise writing concise and precise answers.

Figure 25.4 Preparing for Section B

A summary of what you need to know and do

REVISED ☐

Suitable question for geographical enquiry

- The factors that need to be considered when selecting suitable questions/hypotheses for geographical enquiry.
- The geographical theory/concept underpinning the enquiry.
- Appropriate sources of primary and secondary evidence, including locations for fieldwork.
- The potential risks of both human and physical fieldwork and how these risks might be reduced.

Selecting, measuring and recording data appropriate to the chosen enquiry

- Difference between primary and secondary data.
- Identification and selection of appropriate physical and human data.
- Measuring and recording data using different sampling methods.
- Description and justification of data collection methods.

Selecting appropriate ways of processing and presenting fieldwork data

- Appreciation that a range of visual, graphical and cartographic methods is available.
- Selection and accurate use of appropriate presentation methods.
- Description, explanation and adaptation of presentation methods.

Describing, analysing and explaining fieldwork data

- Description, analysis and explanation of the results of fieldwork data.
- Establish links between data sets.
- Use appropriate statistical techniques.
- Identification of anomalies in fieldwork data.

Reaching conclusions

- Draw evidenced conclusions in relation to original aims of the enquiry.

Evaluation of geographical enquiry

- Identification of problems of data collection methods.
- Identification of limitations of data collected.
- Suggestions for other data that might be useful.
- Extent to which conclusions were reliable.

Unit 3 success checklist

☐ Make the most of the resource booklet – make sure you understand everything, can make use of data and can recognise trends and patterns.

☐ Be very strict with timing – you must not exceed 37 minutes for Section A.

☐ Be concise and precise in your answers – avoid waffle.

☐ Refer to the stimulus material in the exam paper, using facts and figures to support your answer.

☐ Be prepared to construct maps and graphs as well as interpret.

☐ Learn your fieldwork enquiries and make sure that you can write the titles.

☐ Focus on the reasons for your decisions and be prepared to justify what you did.

Notes

Now test yourself and Exam practice answers at **www.hoddereducation.co.uk/myrevisionnotes**